*Alan C. Cassells, PhD*
*Peter B. Gahan, PhD*

# Dictionary of Plant Tissue Culture

*Pre-publication*
*REVIEWS,*
*COMMENTARIES,*
*EVALUATIONS . . .*

"This book is a compendium of the technical terms in the field of plant tissue culture. One of the main features that distinguishes this book from the standard biology dictionaries is that many of the key concepts of the trade are illustrated with clear figures. Another nice feature is the list of key references for the major entries. It will be a handy book for all new practitioners of tissue culture, and will become a must-have book for all plant tissue culture labs."

**Prakash P. Kumar, PhD**
Associate Professor,
Department of Biological Sciences,
National University of Singapore;
Reviews Editor, *Plant Cell Reports*

"The world of plant tissue culture is expanding rapidly and new advances in the related areas of plant biotechnology continue to add new terms and definitions regularly. This dictionary of plant tissue culture is a monumental work that covers the field comprehensively. It provides the reader with a clear, concise, and broad overview of the principles and technology of plant tissue culture. I very much liked the self-explanatory flow charts, tables, and illustrations. These can be effectively used as teaching material and will be greatly appreciated by teachers at all levels. Students in general plant biology and biotechnology programs will also find this resource quite useful. I would highly recommend this dictionary as a resource to students, researchers, teachers, and those who are simply curious about the fascinating world of plant tissue culture."

**Praveen K. Saxena, PhD**
Professor, Department of Plant Agriculture,
University of Guelph

# Dictionary of Plant Tissue Culture

# FOOD PRODUCTS PRESS®
## Crop Science
### Amarjit S. Basra, PhD
### Editor in Chief

*The Lowland Maya Area: Three Millennia at the Human-Wildland Interface* edited by A. Gómez-Pompa, M. F. Allen, S. Fedick, and J. J. Jiménez-Osornio

*Biodiversity and Pest Management in Agroecosystems, Second Edition* by Miguel A. Altieri and Clara I. Nicholls

*Plant-Derived Antimycotics: Current Trends and Future Prospects* edited by Mahendra Rai and Donatella Mares

*Concise Encyclopedia of Temperate Tree Fruit* edited by Tara Auxt Baugher and Suman Singha

*Landscape Agroecology* by Paul A. Wojtkowski

*Concise Encyclopedia of Plant Pathology* by P. Vidhyasekaran

*Molecular Genetics and Breeding of Forest Trees* edited by Sandeep Kumar and Matthias Fladung

*Testing of Genetically Modified Organisms in Foods* edited by Farid E. Ahmed

*Fungal Disease Resistance in Plants: Biochemistry, Molecular Biology, and Genetic Engineering* edited by Zamir K. Punja

*Plant Functional Genomics* edited by Dario Leister

*Immunology in Plant Health and Its Impact on Food Safety* by P. Narayanasamy

*Abiotic Stresses: Plant Resistance Through Breeding and Molecular Approaches* edited by M. Ashraf and P. J. C. Harris

*Teaching in the Sciences: Learner-Centered Approaches* edited by Catherine McLoughlin and Acram Taji

*Handbook of Industrial Crops* edited by V. L. Chopra and K. V. Peter

*Durum Wheat Breeding: Current Approaches and Future Strategies* edited by Conxita Royo, Miloudi M. Nachit, Natale Di Fonzo, José Luis Araus, Wolfgang H. Pfeiffer, and Gustavo A. Slafer

*Handbook of Statistics for Teaching and Research in Plant and Crop Science* by Usha Rani Palaniswamy and Kodiveri Muniyappa Palaniswamy

*Handbook of Microbial Fertilizers* edited by M. K. Rai

*Eating and Healing: Traditional Food As Medicine* edited by Andrea Pieroni and Lisa Leimar Price

*Physiology of Crop Production* by N. K. Fageria, V. C. Baligar, and R. B. Clark

*Plant Conservation Genetics* edited by Robert J. Henry

*Introduction to Fruit Crops* by Mark Rieger

*Generations Gardening Together: Sourcebook for Intergenerational Therapeutic Horticulture* by Jean M. Larson and Mary Hockenberry Meyer

*Agriculture Sustainability: Principles, Processes, and Prospects* by Saroja Raman

*Introduction to Agroecology: Principles and Practice* by Paul A. Wojtkowski

*Handbook of Molecular Technologies in Crop Disease Management* by P. Vidhyasekaran

*Handbook of Precision Agriculture: Principles and Applications* edited by Ancha Srinivasan

*Dictionary of Plant Tissue Culture* by Alan C. Cassells and Peter B. Gahan

*Handbook of Potato Production, Improvement, and Postharvest Management* edited by Jai Gopal and S. M. Paul Khurana

# Dictionary of Plant Tissue Culture

Alan C. Cassells, PhD
Peter B. Gahan, PhD

Food Products Press®
An Imprint of The Haworth Press, Inc.
New York • London • Oxford

For more information on this book or to order, visit
http://www.haworthpress.com/store/product.asp?sku=5648

or call 1-800-HAWORTH (800-429-6784) in the United States and Canada
or (607) 722-5857 outside the United States and Canada

or contact orders@HaworthPress.com

Published by

Food Products Press®, an imprint of The Haworth Press, Inc., 10 Alice Street, Binghamton, NY 13904-1580.

Cover design by Jennifer M. Gaska.

Cover photograph courtesy of Alan C. Cassells.

### Library of Congress Cataloging-in-Publication Data

Cassells, A. C.
    Dictionary of plant tissue culture / Alan C. Cassells, Peter B. Gahan.
        p. cm.
    Includes bibliographical references and index.
    ISBN-13: 978-1-56022-918-6 (hard : alk. paper)
    ISBN-10: 1-56022-918-7 (hard : alk. paper)
    ISBN-13: 978-1-56022-919-3 (soft : alk. paper)
    ISBN-10: 1-56022-919-5 (soft : alk. paper)
    1. Plant tissue culture—Dictionaries. I. Gahan, Peter B. II. Title.

QK725.C34 2006
571.5'38203—dc22

2005030565

# CONTENTS

# ABOUT THE AUTHORS

**Alan C. Cassells** is Professor of Botany at The National University of Ireland, Cork, and has more than 30 years of research and lecturing experience in plant science. He was managing director of a commercial campus plant biotechnology company for 10 years and acts as a consultant to multinational companies and to the International Atomic Energy Agency. He has served on the management committees of several EU COST actions on tissue culture and mycorrhizal fungi. Dr. Cassells is the author of 150 research papers, and editor or co-editor of six books on plant tissue culture. He is a named inventor on four patents. Dr. Cassells has a BSc and MSc from The National University of Ireland and a PhD from the University of Wales, Swansea, where he studied in the department of Professor H. E. Street, a pioneer of plant tissue culture. He was a junior research fellow in Wolfson College, University of Oxford, and lectured at Wye College, University of London. He is a member of the Royal Irish Academy.

**Peter B. Gahan,** formerly Professor of Botany and Dean of Science at Queen Elizabeth College and Professor of Botany and assistant principal of King's College London in the University of London, is currently Emeritus Professor of Cell Biology at KCL and on the management committee of COST action 843. He has taught and researched in the United Kingdom, Canada, France, Italy, and Switzerland and has published three books and some 250 reviews and research papers in international journals. His research has involved the application of quantitative cytochemistry to the study of developmental plant biology; he was the first to demonstrate the lipid components of chromatin and among the first to demonstrate the cytoplasmic location of DNA. He is currently researching the basis of recalcitrance and the competence of calluses and cells to form roots and shoots and the role(s) of circulating DNA in higher plants.

# PREFACE AND ACKNOWLEDGMENTS

Similar to other fields of scientific research, plant tissue culture has developed its own technical terms. The value of a dictionary such as this lies in providing definitions of these terms, thereby helping the reader understand the literature of the field. It is anticipated that this dictionary will find its readership among teachers, researchers, and undergraduate and postgraduate students in basic and applied plant tissue culture. These include people involved in tissue culture per se and also those using plant tissue culture systems for plant cloning (micropropagation), plant secondary metabolite production, plant pathology, and genetic manipulations ranging from pollen culture and somatic hybridization to mutation breeding and genetic engineering.

Most entries include literature citations, and key concepts are illustrated. We have not provided chemical structures as these are readily available on the Internet from sites such as the "Compendium of Pesticide Common Names," which provides extensive information on plant growth regulators (www.hclrss.demon.co.uk/index.html).

The authors are grateful to Dr. Barbara Doyle for proofreading the text. It is always difficult to set boundaries to a technical dictionary and we would appreciate feedback from readers, who are invited to send their comments and suggestion for further entries for the next edition to tcdictionary@ucc.ie.

doi:10.1300/5648_b

# USER'S GUIDE

Entries, including abbreviations and acronyms, have been arranged in alphabetical order, and most include reference citations. The citations point mainly to general textbooks in the core background areas of plant anatomy, biochemistry and histochemistry, developmental biology, genetics, microbiology, micropropagation, plant breeding, plant biotechnology, plant pathology, and plant tissue culture. The many references to papers and reviews on core topics are an introduction to the topic and are not intended to be comprehensive.

Entries are cross-referenced within the text where appropriate.

Key journals in plant tissue culture are *Plant Cell Reports, Plant Cell Tissue and Organ Culture,* and *In Vitro Cellular and Developmental Biology—Plant.* The topic is also covered extensively in horticulture and botanical journals. An earlier dictionary, *Glossary of Plant Tissue Culture* (D. J. Donnelly and W. E Vidaver, 1988), was published by Dioscorides Press of Portland, Oregon.

doi:10.1300/5648_c

**ABA:** *See* ABSCISIC ACID.

**abaxial:** Describes the side of the leaf facing away from the stem. *See* ADAXIAL.

**abiotic factors:** Environmental factors that influence plant growth and survival. These include light, oxygen, carbon dioxide, water, minerals, and temperature (Lerner, 1999). *See* BIOTIC FACTORS.

**abiotic stress:** Stress as a result of environmental as opposed to biotic factors (see Figure 1). Common environmental stresses include drought stress, waterlogging, temperature stress, light stress, and salt stress (Taiz and Zeigler, 2002). Some of these stresses may also be experienced by tissues in vitro (Cassells and Curry, 2001; Gaspar et al., 2002; Joyce et al., 2003). *See* BIOTIC STRESS; HYPERHYDRICITY.

**abscisic acid:** ABA; a sesquiterpenoid synthesized from carotinoids in the chloroplasts and other plastids. Its production is associated with water stress and inhibition of shoot but not root growth. ABA is involved in the induction and maintenance of dormancy and in the expression of proteinase inhibitors in the response to wounding. It is transported in the phloem and xylem (Hopkins and Hunter, 2002; Taiz and Zeigler, 2002). *See* ARTIFICIAL SEED.

**abscission:** The programmed shedding of plant organs, e.g., leaves, fruits, flowers. Several plant hormones are implicated in the process. It was originally believed that ABA controlled leaf abscission, but this is the case in only a few species. Ethylene accelerates abscission. Auxin acts as a suppressor of abscission but becomes a promoter at supraoptimal concentration, where it stimulates ethylene production. Hence both ethylene-releasing compounds and auxins are used commercially as defoliants. ABA affects leaf senescence, possibly by promoting ethylene synthesis (Srivastava, 2002). *See* ABSCISIC ACID; DEFOLIANTS.

*1*

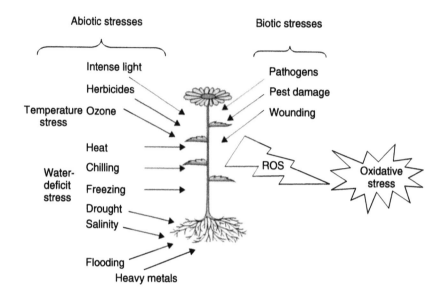

FIGURE 1. Illustration of the causes of abiotic and biotic stresses (Slater et al., 2003). Reprinted by permission. All stresses induce a general oxidative stress and more specific stress responses (Lerner, 1999; Taiz and Zeigler, 2002). ROS: reactive oxygen species. *See* ABIOTIC STRESS; BIOTIC STRESS; OXIDATIVE STRESS.

**abscission layer:** The target cells in the abscission zone, which are sensitized to ethylene by reduction of auxin and in which ethylene activates cell-wall-degrading enzymes (Taiz and Zeigler, 2002).

**absorption spectrum:** The amount of energy absorbed by a substance, e.g., a pigment, plotted against the wavelength of the light (Taiz and Zeigler, 2002). *See* ACTION SPECTRUM; GROWTHROOM LIGHTING; PHOTOSYNTHESIS.

**ACC:** *See* AMINOCYCLOPROPANECARBOXYLIC ACID.

**acclimation:** *See* ACCLIMATIZATION.

**acclimatization:** The physiological process of adaptation of an organism to function more efficiently in changed environmental circumstances. It is usually short term and reversible (Lerner, 1999). Acclimatization is a critical stage (Stage 4) in micropropagation (Preece and Sutter, 1991; George, 1996). *See* MICROPROPAGATION.

**acetocarmine:** A dye used to stain chromosomes, now largely replaced by Giemsa staining (Gahan, 1984). *See* CHROMOSOME; GIEMSA.

**aceto-orcein:** A dye used to stain chromosomes, now largely replaced by Giemsa staining (Gahan, 1984). *See* CHROMOSOME; GIEMSA.

**achromycin:** A tetracycline antibiotic (Walsh, 2003). *See* ANTIBIOTICS.

**acropetal:** Directed from the base toward the apex of the organ or plant. Water and chemicals generally move acropetally. Development occurs acropetally. *See* BASIPETAL.

**actinomycetes:** A major subdivision of the prokaryotes. Actinomycetes are sometimes described as the "higher bacteria," possessing properties intermediate between the fungi and the bacteria. The actinomycetes are gram-positive organisms that grow slowly as branching filaments, resembling fungi, as their filamentous growth forms mycelial colonies. The actinomycetes are divided into several families. The Actinomycetaceae family consists of two genera: *Actinomyces* and *Nocardia;* the Mycobacteriaceae family has a single genus, *Mycobacterium;* the Streptomycetaceae family comprises soil-inhabiting organisms, some of which produce antibiotics (Ortiz-Ortiz et al., 1984). *See* ANTIBIOTICS; PROKARYOTE.

**actinomycin D:** An antibiotic that functions as a transcriptional terminator and acts by binding to DNA. The tricyclic ring system (phenoxazone) intercalates between adjacent G-C base pairs, and the cyclic polypeptide arms fill the nearby cavity (Walsh, 2003). *See* ANTIBIOTICS.

**action spectrum:** A plot of the relative rate of a process as affected by different light wavelengths. *See* ABSORPTION SPECTRUM; PHOTOSYNTHESIS.

**activated charcoal:** Charcoal that has been activated by heat or acid washing. It is added to plant tissue culture as a nonselective agent for binding aromatic and olefinic compounds, e.g., ethylene. Charcoal may also bind/release ions and darkens the root environment, thereby facilitating light-sensitive reactions (George, 1993).

**adaptation:** Inherited stress resistance (Lerner, 1999). *See* ACCLIMA-TIZATION.

**adaxial:** Describes the side of the leaf facing toward the stem. *See* ABAXIAL.

**adenine:** Syn. vitamin B4; a purine base found in nucleic acids, in energy donors (e.g., adenosine triphosphate), and in natural cytokinins (Dey and Harborne, 1996). *See* CYTOKININS.

**adenine sulfate:** A weak cytokinin, reported to be beneficial when it is added to a cytokinin-containing medium (George, 1996). *See* CYTOKININS.

**adenosine triphosphate:** A high-energy phosphate ester that is the main energy storage compound of the cell (Metzler, 2001).

**adult phase:** Plant development involves several more or less discrete stages, which are termed phase changes. These stages are expressed in terms of morphological, anatomical, and physiological traits, which change over time, reflecting the juvenile–adult phase change (Poethig, 1990; Howell, 1998). Cuttings from the adult phases of plants may be difficult to root. Buds from these tissues can be "rejuvenated/reinvigorated" by culture in vitro, usually by micrografting to seedling rootstocks (George, 1993). *See* HETEROBLASTY; HETEROPHYLLY; MICROGRAFTING; PHASE CHANGE; REINVIGORATION; REJUVENATION.

**adult plant:** A plant that is developmentally mature enough to flower. *See* JUVENILE PHASE.

**adult plant resistance:** Polygenic resistance to biotic and abiotic stresses expressed in mature, as opposed to young, plant tissues

(Allard, 1999; Strange, 2003). *See* DISEASE RESISTANCE; MAJOR GENE RESISTANCE.

**adventitious buds:** Buds that arise from tissues lacking preformed buds, e.g., from callus (George, 1993).

**adventitious regeneration:** The development of roots or shoots from tissues lacking preformed buds. These organs may arise directly from cells of the explant (direct organogenesis) or from callus arising on or subcultured from the explant (indirect organogenesis). Such shoots, particularly those arising via indirect organogenesis, may show genetic variability ("somaclonal variation") (George, 1993; Jayasankar, 2005). *See* SOMACLONAL VARIATION.

**adventitious roots:** Also cladogenous roots; roots arising from aerial parts, underground stems, and relatively old roots, e.g., from the hypocotyls of gymnosperms and stem nodes of monocots. May have several orders of branching but lack secondary growth (Esau, 1977; Mauseth, 1988). *See* RHIZOGENESIS.

**adventitious shoots:** *See* ADVENTITIOUS REGENERATION; CAULOGENESIS.

**aeration:** Provision of oxygen under pressure as opposed to ambient air supply; used in liquid cultures to prevent anoxia (Hvoslef-Eide and Preil, 2004). *See* ANOXIA.

**aerenchyma:** Large gas-filled spaces in the roots associated with hypoxia (Esau, 1977; Mauseth, 1988).

**aeroponics:** A form of hydroponic culture that involves misting the plant's tissues, normally roots, with nutrient solution. The runoff is collected and can be analyzed and supplemented and recycled as in nutrient film technique (NFT) systems. Some systems mist the tissues in timed cycles, whereas others constantly mist the roots. One of the advantages of aeroponics, such as NFT systems, is that the medium can be modified without transfer of the tissues (Sholto Douglas, 1986; Winsor and Schwarz, 1990; Schwarz, 1995). *See* HYDROPONICS.

**AFLP:** *See* AMPLIFIED FRAGMENT LENGTH POLYMORPHISM TECHNIQUE.

**agar:** A polysaccharide that forms a solid gel on cooling after solubilization. It is the gelling agent most widely used to support plant tissues in culture to prevent anoxia as a result of submersion in liquid media. Agar is prepared from red algae *(Rhodophyta),* from species of *Gelidium* (mainly) and *Gracilaria,* and consists of a mixture of agarose and agaropectin. Agar, a natural product, is variable in quality, and commercial agars may also be derived from different algal species in different seasons; accordingly, individual batches of commercial agars should be tested for quality (Cassells and Collins, 2000). The quality of agar is improved by alkaline treatment that converts any L-galactose-6-sulfate to 3,6-anhydro-L-galactose (Bryant et al., 1999; Imeson, 1999). *See* AGAROPECTIN; AGAROSE; GELLING AGENT.

**agaropectin:** A component of agar consisting of a heterogeneous mixture of molecules of lower molecular weight than agarose that occur in lesser amounts. The structures of agarose and agaropectin are similar but the latter is slightly branched and sulfated, and they may have methyl and pyruvic acid ketal substituents. They gel poorly and may be removed from agarose molecules by charge separation (Imeson, 1999). *See* AGAR; AGAROSE.

**agarose:** A component of agar that is a linear polymer of molecular weight ~120,000 kDa based mainly on D-galactose units and the 3,6-anhydro form of L-galactose with small amounts of D-xylose. Some of the D-galactose units are methylated. *See* AGAR; GELLING AGENT.

**agglutination:** Clumping of particles by antibodies (Coico et al., 2003). A technique used for detecting and quantifying antigens and antibodies used, e.g., in virology (Dijkstra and de Jager, 1998). *See* ANTIBODY; ANTIGEN.

***Agrobacterium*-mediated gene transfer:** Involves gene transfer via a modified tumor-inducing (Ti) plasmid. Although not suitable for all species, it is a widely used method that has the advantage of inducing less rearrangement of the transgene and lower transgene copy number than in direct gene transfer methods (Slater et al., 2003; Li and Gray, 2005). *See AGROBACTERIUM TUMEFACIENS;* GENETIC ENGINEERING; GENE TRANSFER METHODS.

***Agrobacterium rhizogenes:*** A gram-negative, motile, aerobic bacterium that may contain the root-inducing (Ri) plasmid, which inserts into the chromosomes of the host plant cell. The Ri plasmid, similar to the Ti plasmid, contains a strong promoter and genes encoding the production of indole-3-acetic acid and various cytokinins (Sigee, 1993). This bacterium transfers its transfer DNA (T-DNA), which is a portion of the Ri plasmid, to susceptible plant cells, where the T-DNA, if integrated into the nuclear genome of the plant cell, encodes genes that direct the synthesis of auxin (indole-3-acetic acid), increase the sensitivity of the transformed plant cells to auxin, or both. The endogenous production of auxin or an increase in auxin sensitivity can lead to the formation of adventitious roots at the site of infection (Sigee, 1993). *Agrobacterium rhizogenes* has been used to enhance adventitious root formation in vitro. *See AGROBACTERIUM TUMEFACIENS; ROL* GENES; *ROOTING.*

***Agrobacterium tumefaciens:*** A gram-negative, motile, aerobic bacterium that may contain the tumor-inducing (Ti) plasmid, which inserts into the chromosomes of the host plant cells. The Ti plasmid contains a strong promoter and genes encoding the production of indole-3-acetic acid and various cytokinins (Sigee, 1993). The transfer DNA (T-DNA), a specific segment of the Ti plasmid, is stably incorporated into the host genome, resulting in cell division and proliferation giving rise to crown gall disease (Agrios, 1997). By excising the oncogenes and replacing them with a novel gene (e.g., one coding for a product conferring resistance to a plant pathogen), genetically transformed plants can be produced. The plasmid usually contains a selectable marker gene (e.g., resistance to the antibiotic kanamycin) and may also contain a reporter gene *(GUS).* In this example, the former allows only transformed plants to survive in a kanamycin-containing selection medium, and the *GUS* gene gives a phenotype producing a blue color on reacting with histochemical reagents, allowing easy detection of transformed plants (Chawla,

2002; Slater et al., 2003). *See* β-GALACTURONIDASE GENE; GENETIC ENGINEERING; GENE TRANSFER METHODS; SELECTABLE MARKER.

***Agrobacterium tumefasciens:*** *See AGROBACTERIUM TUMEFACIENS.*

**air filtration:** The removal of particles and gases from air. In tissue culture this generally refers to the use of filters to remove bacterial and fungal spores from the air supply in laminar flow/clean air cabinets. *See* HEPA FILTER.

**airlift bioreactor:** A culture vessel in which forced air is used to maintain the tissue in suspension (Scragg, 1991a; Hvoslef-Eide and Preil, 2004). *See* BIOREACTOR.

**airlift fermenter:** *See* AIRLIFT BIOREACTOR.

**ALAR:** *See* AMINOCYCLOPROPANECARBOXYLIC ACID.

**albino plants:** Plants lacking pigment; generally refers to plants lacking chlorophyll for genetic or environmental reasons (lack of light).

**alcohol burners:** Burners used for heat sterilization of instruments. *See* INSTRUMENT STERILIZATION.

**alginate:** A carbohydrate produced by brown seaweeds *(Phaeophyceae),* mainly *Laminaria* spp. Alginates are linear unbranched polymers containing D-mannuronic and L-guluronic acid residues (Bryant et al., 1999; Imeson, 1999).

**alkaloids:** A large family of nitrogen-containing plant secondary metabolites frequently produced in response to abiotic and biotic stresses (Roberts and Wink, 1998). *See* ABIOTIC STRESS; BIOTIC STRESS.

**alleles:** Variants of a gene differing in nucleotide sequences (Russell, 2002).

**allelopathy:** The release of plant secondary metabolites that have harmful effects on neighboring plants (Weston and Duke, 2003; Schulz, 2004). *See* SEMIOCHEMICALS.

**allopolyploids:** Polyploids resulting from the crossing of different species that may be sterile (Allard, 1999). Spontaneous or chemically induced chromosome doubling in vivo or in vitro is used to restore their fertility. Chemical antimitotic agents include colchicines, amiprophos-methyl, oryzalin, and trifluralin (Taji et al., 2001; Cassells, 2003; Grzebelus and Adamus, 2004). *See* COLCHICINE; ORYZALIN.

**allotetraploids:** Sterile hybrids resulting from the hybridization of species where the gametes are unreduced diploids. Occasionally diploid gametes are formed that on selfing give fertile tetraploids (Allard, 1999). *See* ALLOPOLYPLOIDS.

**α-particle:** Positively charged helium nucleus emitted at high velocity during radioactive decay.

**α-tocopherol:** Vitamin E; lipid-soluble terpenoid-like compound found in high concentrations especially in particular seeds, e.g., cereals. Vitamin E appears to prevent lipid oxidation and can replace the effects of vernalization.

**α-tocopheryl acetate:** Form of tocopherol added to plant tissue culture medium.

**AMF:** *See* ARBUSCULAR MYCORRHIZAL FUNGI.

**amino acids:** Molecules containing amino ($-NH_2$) and carboxylic acid ($-COOH$) groups with the formula $RCHNH_2COOH$, where R varies from hydrogen to aromatic and heterocyclic rings. Amino acids are the basic units of proteins, although many amino acids are known that are not found in proteins. Individual amino acids such as glycine and hydrolyzed proteins, e.g., casein hydrolysate, are added to plant tissue culture media (Wallsgrove, 1995). *See* CASEIN HYDROLYSATE; PLANT TISSUE CULTURE MEDIA.

**aminocyclopropanecarboxylic acid:** 1-Aminocyclopropanecarboxylic acid; the immediate precursor compound from which ethylene is synthesized (Srivastava, 2002; Taji and Zeigler, 2002). *See* ETHYLENE.

**aminoethoxyvinylglycine:** AVG; an inhibitor of ethylene biosynthesis; inhibits the conversion of *S*-adenosylmethionine to 1-aminocyclopropanecarboxylic acid (Taiz and Zeigler, 2002). *See* AMINOOXYACETIC ACID; ETHYLENE INHIBITORS.

**aminooxyacetic acid:** AOA; an inhibitor of ethylene synthesis that blocks the conversion of *S*-adenosylmethionine to 1-aminocyclopropanecarboxylic acid (Taiz and Zeigler, 2002). *See* AMINOCYCLOPROPANECARBOXYLIC ACID; AMINOETHOXYVINYLGLYCINE; ETHYLENE INHIBITORS.

**aminopenicillin:** *See* AMPICILLIN.

**AMO 1618:** A commercial antitranspirant (Basra, 2000). *See* ANTITRANSPIRANTS.

**amphidiploid:** An allopolyploid that behaves similarly to a diploid at meiosis; i.e., it forms only bivalents (Grant, 1975). *See* ALLOPOLYPLOIDS; BIVALENTS.

**ampicillin:** A penicillin; unlike penicillin, ampicillin and the closely related amoxicillin can penetrate and prevent the growth of gramnegative bacteria (Walsh, 2003). *See* ANTIBIOTICS.

**amplified fragment length polymorphism technique:** AFLP technique; a method for fingerprinting genotypes based on selective amplification of a subset of restriction fragments from the cutting of genomic DNA with endonucleases. Polymorphisms are detected using gel electrophoresis based on differences in length of the amplified fragments (Matthes et al., 1998). *See* GENETIC FINGERPRINTING; POLYMERASE CHAIN REACTION.

**amylopectin:** A branched polymer of glucose (Bryant et al., 1999). *See* STARCH.

**amyloplast:** Nonpigmented plastid or leucoplast synthesizing starch (Graham et al., 2003). *See* ORGANELLE.

**amylose:** An unbranched polymer of glucose (Bryant et al., 1999). *See* STARCH.

**analysis of variance:** ANOVA; a statistical analysis of how much of the variability in a set of observations can be ascribed to different causes (Motulsky, 1995).

**anaphase:** *See* MEIOSIS; MITOSIS.

**anatomy:** Plant structure or the study of plant cells and tissues (Mauseth, 1988; Esau, 1977). Aspects relevant to plant tissue culture are discussed in Trigiano and Gray (2005).

**ancymidol:** An inhibitor of gibberellin synthesis that acts at the conversion of kaurene to kaurenoic acid (Basra, 2000). *See* GIBBERELLIN INHIBITORS.

**androgenesis:** Development from a male gamete (Taji et al., 2001; Reed, 2005).

**aneuploid:** A cell, tissue, or plant in which the chromosome number is not an exact multiple of the haploid chromosome number (Allard, 1999).

**Angiospermae:** Flowering plants with ovules enclosed in carpels.

**angiosperms:** Flowering vascular plants, all bearing seeds within enclosed carpels (Heywood, 1993; Willis and McElwain, 2002).

**annual plant:** A plant that completes its life cycle in a year or less (Hartmann et al., 2001).

**ANOVA:** *See* ANALYSIS OF VARIANCE.

**anoxia:** Absence of oxygen. *See* HYPOXIA.

**anther:** Stamen tissue that produces pollen (microspores) (Esau, 1977).

**anther culture:** The culture of anthers to produce haploid plants (see Figure 2). Haploids can be chemically induced to form homozygous diploids or may undergo spontaneous endoduplication (Taji et al., 2001; Reed, 2005).

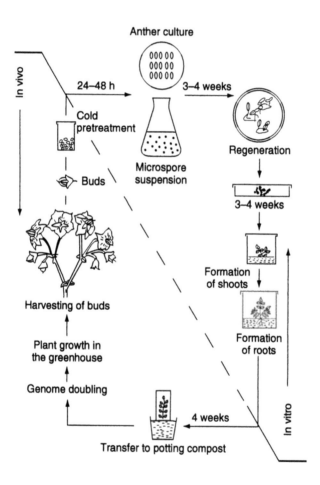

FIGURE 2. The stages in anther culture; note the cold pretreatment of the buds (Foroughi-Wehr and Wenzel, 1994). Reprinted by permission of Springer Science and Business Media. *See* ANTHER CULTURE; POLLEN CULTURE.

**anthocyanins:** Pigmented, glycosylated flavanoids that constitute many of the blue, purple, pink, and red pigments in plants (Hendry, 1993). *See* FLAVANOIDS.

**antiauxin:** *See* AUXIN INHIBITORS.

**antibacterials:** *See* ANTIBIOTICS.

**antibiotic resistance:** Resistance to an antibiotic, rendering it no longer effective. Resistance may be based on either destruction or modification of the antibiotic by the target organism and by either altered uptake by the target organism or changes in the antibiotic target. Multiple antibiotic resistance also occurs (Walsh, 2003). *See* ANTIBIOTICS.

**antibiotics:** Drugs used to treat infections caused by bacteria; the term is sometimes used in human therapy to include antifungal compounds (Walsh, 2003; Exhibit 1). Antibiotics, i.e., antibacterials, are used in some countries, including the United States, to treat field diseases, but in many countries their use in the field is prohibited. Antibiotics are commonly used in plant tissue culture (Barrett and Cassells, 1994) or applied to the stock plant (Cassells et al., 1988). *See* ANTIMYCOTICS; ANTIVIRAL COMPOUNDS.

**antibody:** An immunoglobulin (protein) produced by blood plasma cells on exposure to an antigen that specifically recognizes and binds to that antigen. Many antigens have more than one recognition site (determinant), and so blood serum contains polyclonal antibodies, i.e., mixtures of antibodies to the respective antigen determinants (Roitt et al., 2001). Antibodies form the basis of serological tests to detect and identify pathogens (Dijkstra and de Jager, 1998; Martin, 1998). *See* ENZYME-LINKED IMMUNOSORBENT ASSAY; LATEX AGGLUTINATION; MONOCLONAL ANTIBODY.

**anticlinal cell wall:** A wall perpendicular to the surface of the cell. *See* APICAL MERISTEM.

---

### EXHIBIT 1. Antibiotics: antibacterial compounds

*Origin:* First isolated (first-generation antibiotics) from *Penicillium* spp.; subsequently mainly isolated from *Bacillus, Streptomyces,* and *Cephalosporium* spp. Many derivatives (second- and third-generation antibiotics) are synthetic. Purely synthetic antibiotics include sulphonamides and quinolones.

*Characteristics:* Antibiotics are either narrow spectrum (limited effectiveness) or broad spectrum (affect a wide range of bacteria). They can be bacteriocidal (kill the target bacteria) or bacteriostatic (prevent bacterial multiplication).

*Mode of action:* Antibiotics can inhibit

- Nucleic acid metabolism, e.g., sulphonamides, trimethoprim, sulphonamides, rifampicin, quinolones
- Protein synthesis, e.g., aminoglycosides (streptomycin, gentamicin, tobramycin), chloramphenicol, macrolides (erythromycin), tetracycline
- Cell wall synthesis, e.g., bactricin, cephalosporins (cephotaxime, ceftriaxone), penicillins (methacillin, amphicillin, amoxacillin), vancomycin
- Membrane function, e.g., aminogycosides, imidaxoles, polyene, polymixin

*Selection of antibiotic for in vitro use (Cassells and Barrett, 1994):*

- Isolate the bacterial contaminant in pure culture.
- Determine antibiotic sensitivity on drug sensitivity testing medium.
- Determine the minimum inhibitory concentration (MIC)—the minimum amount necessary to inhibit growth of the bacterium.
- Determine the minimum bacteriocidal concentration (MBC)—the amount necessary to kill the bacterium. Antibiotics are conventionally used at the MBC × 4.
- Screen the antibiotic for plant culture toxicity at the MBC × 4.
- If nontoxic to the plant tissue, culture contaminated tissue on MBC × 4.

*Reference:* Walsh, C. (2003). *Antibiotics: Actions, Origins, Resistance.* Herndon, VA: ASM Press.
*Note:* Some antibiotics, e.g. cefotaxime can promote adventitious regeneration.

---

**anticlinal division:** Cell division where the cell wall forms at a right angle to the surface, thereby increasing the girth of the tissue (Esau, 1977). *See* APICAL MERISTEM.

**anticytokinin:** *See* CYTOKININ INHIBITORS.

**antifungal compounds:** *See* ANTIMYCOTICS; FUNGICIDES.

**antigen:** Any molecule capable of inducing an immune response, i.e., capable of inducing antibodies (Roitt et al., 2001). Small molecules incapable of inducing an immune response, e.g., plant growth regulators, can be covalently linked to proteins to induce antibody formation. Commercial kits are available for the detection of pathogens and plant growth regulators based on specific antibodies. *See* ANTIBODY.

**antigibberellin:** *See* GIBBERELLIN INHIBITORS.

**antimycotics:** Chemicals that inhibit fungi. Chemicals developed for plant fungal disease control are generally referred to as fungicides; the term antimycotics is applied to fungicides developed for human or animal use. The latter are sometimes used in plant tissue culture. Generally, the active compounds of fungicide formulations are used in tissue culture. Human antifungal compounds, e.g., amphotericin B, griseofulvin, and nystatin, have been used in tissue culture (George, 1993). *See* BENOMYL; FUNGICIDES.

**antioxidant:** Antioxidants react with and inactivate free radicals (Scott, 1997). *See* FREE RADICALS; OXIDATIVE STRESS.

**antisense mRNA:** Messenger RNA transcribed from a cloned gene the sequence of which is complementary to the mRNA produced by the normal gene. This could be complementary to that of a virus coat protein gene, for instance, thus affording virus resistance (Slater et al., 2003). *See* GENETIC ENGINEERING.

**antiserum:** Serum containing antibodies (Roitt et al., 2001). *See* ANTIBODY.

**antitranspirants:** Compounds that cause stomata to close or coat the plant surface and thereby reduce water loss from the plant (Basra, 2000). Antitranspirants are used in the field to reduce stress and have been used to protect microplants against water stress during establishment. *See* DROUGHT; STOMA; WATER STRESS.

**antiviral compounds:** Compounds that inhibit viral replication. A number of antiviral compounds are used in human therapy; of these, ribavirin (trade name: Virazole) has been used in plant tissue culture to eliminate viruses (Cassells and Long, 1980a,b; Cassells, 1983; Hull, 2002). *See* RIBAVIRIN.

**AOA:** *See* AMINOOXYACETIC ACID.

**apical:** Describes the tip as opposed to the base.

**apical dominance:** Inhibition of the lateral buds by the apical bud. This can be overcome by the use of either cytokinins applied to the lateral buds in lanolin paste or auxin inhibitors, e.g., triiodobenzoic acid, applied to the apical bud in lanolin paste or by excision of the apical bud (Taiz and Zeigler, 2002). Apical dominance is often lost in in vitro cultures, which may have a tendency toward bushiness (George, 1993).

**apical meristem:** Primary meristem; the area of cell division at the tip of the shoot and root that gives rise to the primary tissues of the plant and is responsible for increased shoot length (Howell, 1998; Lyndon et al., 1998; see Figure 3 also Figures 7 and 8). *See* CAMBIUM; LATERAL MERISTEM.

**apical necrosis:** Death of the tip of the shoot, associated with adverse media and environmental factors in tissue culture (Ziv, 1991; Ziv and Ariel, 1994). *See* HYPERHYDRICITY.

**API kits:** Commercial miniaturized biochemical kits for the identification of environmental bacteria and yeasts. *See* GOOD LABORATORY PRACTICE.

**apomixis:** Usually used to refer to the production of seeds without fertilization but occasionally used to describe any form of asexual or vegetative reproduction (Grant, 1975). *See* VEGETATIVE REPRODUCTION.

**apoplast:** The continuity between cell walls, intercellular spaces, and xylem vessels in plant tissues (Esau, 1977). *See* APOPLASTIC MOVEMENT; ENDOPHYTIC ORGANISMS.

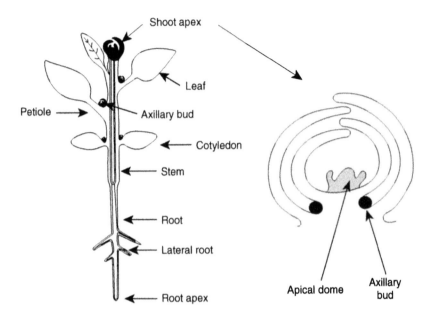

FIGURE 3. Labeled plant showing the tissues commonly used as sources of explants. The preferred explant is the apical dome ("meristem tip") with the first pair of leaf primordia. Larger apical explants may contain axillary buds and increased risk of carrying over microbial contaminants into culture. *See* MERISTEM CULTURE; MICRO-PROPAGATION; VIRUS ELIMINATION.

**apoplastic movement:** The route by which water and water-soluble compounds may move within plant tissues without entering the cell. Bacteria and fungi, lacking the ability to penetrate the plant cell wall, may also colonize the apoplast by surviving on natural or induced cell metabolite leakage. *See* ENDOPHYTIC ORGANISMS; SYMPLASTIC MOVEMENT.

**apoptosis:** A form of programmed cell death in animal cells, initiated by the mitochondria and involving a cascade of caspases, in which nuclear chromatin becomes clumped at the nuclear membrane prior to breakdown of the nucleus and fragmentation of the DNA. No genes and caspases similar to those in animals have been found in plants (Beers and McDowell, 2001; Mittler and Shulaev, 2003). Whereas some forms of programmed plant cell death have been shown to occur in the absence of apoptotic nuclear events, e.g., xylogenesis (Fukuda, 1996; Fath et al., 1999) and barley aleurone (Fath et al., 1999), other systems imply changes that may be accorded the apoptosis label, e.g., "root cortex of soybean root" necrosis mutants (Kosslak et al., 1997), anther development (Wang et al., 1999), aerenchyma formation (Gunawardena et al., 2001), in vitro cell death (McCabe et al., 1997), and cell death involved in the hypersensitive response (Strange, 2003). ABA (Wang et al., 1999), ethylene (Gunawardena et al., 2001), and cytokinins (Gahan et al., 2003) have been implicated in the plant apoptotic process (Danon et al., 2000; Kuriyama and Fukuda, 2002). *See* HYPERSENSITIVITY.

**apospory:** The formation of an embryo from a cell of the ovular tissue where megaspore formation is bypassed (Bhojwani and Soh, 2001). *See* EMBRYOGENESIS; SOMATIC EMBRYOGENESIS.

**arabinose:** A pentose (five-carbon) sugar commonly found in plants (Bryant et al., 1999).

**arbuscular mycorrhizal fungi:** AMF; a group of root-inhabiting symbiotic fungi, characterized by the production of arbuscules, that facilitates phosphate uptake and is considered to protect against pathogen and drought stress (Kapulnik and Douds, 2000; Saxena and Johri, 2002; Duffy and Cassells, 2003). They are used as microplant inoculants in vitro and in vivo, to protect against establishment and weaning stress, and to improve plant field performance (Vestberg et al., 2002). *See* ECTOMYCORRHIZAL FUNGI.

**arbuscules:** Hyphal swellings formed in cells infected by mycorrhizal fungi (Duffy and Cassells, 2003). *See* ARBUSCULAR MYCORRHIZAL FUNGI.

**Archaebacteria:** Syns. Archaea, extremophiles; a division of the prokaryotes with ribosomal RNA related to that of the eukaryotes. Subdivided into thermophiles, which live at high temperatures; hyperthermophiles, which live at very high temperatures; psychrophiles, which are adapted to the cold; halophiles, which live in very saline environments; acidophiles, which live at low pH; and alkaliphiles, which thrive at a high pH (Boone and Castenholz, 1999). *See* EUKARYOTES; PROKARYOTE.

**arginine:** A basic amino acid that is important in histone proteins (Wallsgrove, 1995). *See* HISTONE.

**artificial light:** Light usually provided by tungsten bulbs, fluorescent tubes, sodium or mercury vapor lamps. Most plant cultures are not dependent on light for energy, being provided with a carbon growth source, commonly sucrose. However, they do require light for morphogenesis and to prevent etiolation. It is important when considering lighting for plant cultivation to measure light emission in the photosynthetically active region (PAR) of the spectrum. This is read with a PAR meter and given as photosynthetic photon flux (PFF) in $\mu mol \cdot m^{-1} \cdot s^{-1}$ (George, 1993). A relationship obtains between light intensity and carbon dioxide concentration (Taiz and Zeigler, 2002). At high light intensities, carbon dioxide may limit plant growth in greenhouses and in vitro. Supplementary carbon dioxide may be supplied under these conditions (Hartmann et al., 2001). *See* GROWTHROOM LIGHTING; PHOTOSYNTHESIS.

**artificial seed:** Seed developed from the encapsulation of, usually, a somatic embryo, though other propagules such as nodes of microplants may be used. It usually contains an artificial endosperm and artificial cell wall. The endosperm may consist of a basal medium formulation, e.g., that of Murashige and Skoog (1962), with or without organic components, with or without plant growth regulators, and with or without an energy source, e.g., sucrose. The artificial seed wall may consist of carbohydrates or other biodegradable material (Redenbaugh, 1993; Bajaj, 1995). *See* ENDOSPERM; SEED COAT; SOMATIC EMBRYO.

**ascorbic acid:** Vitamin C; water-soluble vitamin used in tissue culture as an antioxidant (George, 1993). *See* ANTIOXIDANT.

**aseptic conditions:** Sterile conditions; in plant tissue culture achieved usually by autoclaving instruments, culture vessels, and media and equipment, by wiping surfaces with alcohol or other sterilizing solutions, by surface sterilization of plant tissue from the environment with aqueous ethanol or commercial hypochlorite formulation, and by operating in a laminar flow cabinet in a stream of filtered air. *See* AUTOCLAVE; INSTRUMENT STERILIZATION; LAMINAR FLOW CABINET; STERILE FILTRATION; SURFACE STERILANTS.

**aseptic cultures:** Cultures indexed for, and shown to be free of, contamination by cultivable bacteria and yeasts (Cassells, 1997, 2000a; Cassells et al., 2000; Cassells and Doyle, 2004, 2005). *See* AXENIC CULTURE; CULTURE INDEXING; PATHOGEN INDEXING.

**asexual reproduction:** Propagation of plants from vegetative parts such as stem and leaf cuttings, bulbs, corms, tubers, and root cuttings and by grafting and layering (McMillan Browse, 1979; Hartmann et al., 2001). Micropropagation is an alternative to the latter in vivo methods of plant multiplication (Debergh and Zimmerman, 1991; George, 1993, 1996). *See* MICROPROPAGATION; VEGETATIVE PROPAGATION.

**asparagine:** An uncharged polar amino acid formed by the addition of ammonia to aspartic acid. Along with glutamine, it is used to store ammonia for amino acid biosynthesis (Metzler, 2001; Buchanan et al., 2002). Both asparagine and glutamine are occasionally included in plant tissue culture media (George, 1996). *See* PLANT TISSUE CULTURE MEDIA.

**aspartic acid:** An amino acid, constituent of proteins, important in the transfer of amino groups (Metzler, 2001). *See* ASPARAGINE.

**atmosphere:** The gaseous environment that surrounds the earth. It consists of 78.1 percent nitrogen, 20.1 percent oxygen, ~0.5 percent (variable) water vapor, 0.035 percent (350 ppm) carbon dioxide, and trace amounts of volatile organic compounds released by biological and nonbiological processes (Firor, 1992). Some of these volatile organic compounds may reach concentrations at which they can affect plants; e.g., plasticizers in greenhouses have been reported to be toxic to plants. Ethylene, nitric oxide, and other plant stress compounds, e.g., green leaf volatiles, may influence neighboring plants or attract pests, parasitoids, and predators (Strange, 2003; Schulz, 2004; Wendehenne et al., 2004). *See* IN VITRO ATMOSPHERE.

**autoclave:** A vessel in which superheated steam under pressure is used to sterilize tissue culture materials and media. The standard autoclave settings for tissue culture media are 121°C and 1.05 kg·cm$^{-2}$ (103 kPa) (George, 1993). It is important that autoclaves are routinely serviced and checked for cracks in the casing and for correct temperature and pressure conditions. *See* AUTOCLAVE TAPE; GOOD LABORATORY PRACTICE.

**autoclave tape:** An indicator tape applied to material going into the autoclave that indicates the appropriate conditions have been achieved. *See* GOOD LABORATORY PRACTICE.

**automation:** Involves the mechanization of processes aimed at the reduction of production costs. In micropropagation it involves the mechanization of media preparation and sterilization, of tissue subculture, and of transfer to soil or to the field (Vasil, 1991). *See* ARTIFICIAL SEED; MICROPROPAGATION.

**autotrophic culture:** A culture capable of using inorganic sources of carbon (carbon dioxide) and nitrogen (nitrates, ammonium salts) as starting materials for biosynthesis and using sunlight or artificial light as an energy source (Kozai, 1991; Zobayed et al., 2000; see Figure 4). Autotrophic culture is usually performed under aseptic conditions. It may be used to recover infected cultures (Long, 1997). It has similarities to hydroponic culture and is sometimes referred to as microhydroponics (Cassells, 2000b). *See* HETEROTROPHIC GROWTH; MICROHYDROPONICS.

**auxin inhibitors:** Inhibitors of auxin action or movement (Taiz and Zeigler, 2002). Auxin transport is inhibited in explants by naturally occurring compounds, e.g., quercitin and genistein, and by synthetic compounds such as 1-naphthoxyacetic acid (1-NOA), 1-*N*-naphthylamic acid (NPA), and 2,3,5-triiodobenzoic acid (TIBA). Abscisic acid is an antagonist (George, 1993, 1996). *See* PLANT GROWTH REGULATOR.

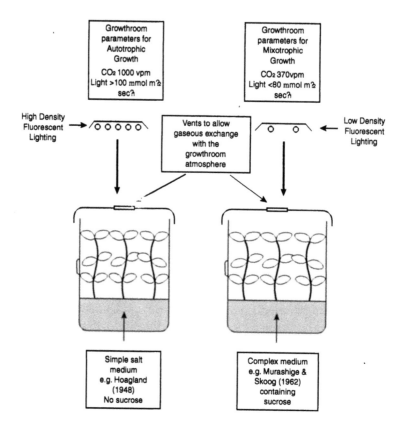

FIGURE 4. A comparison of the light and media composition used in autotrophic culture compared with mixotrophic (see "heterotrophic") culture. Atmospheric carbon dioxide (370 v.p.m. – Volume per million parts of air) limits photosynthesis at high light intensities *in vitro* and must be supplemented to optimize growth. Humidity may be controlled by the use of a humidifier. See AUTOTROPHIC CULTURE; HETEROTROPHIC GROWTH; MIXOTROPHIC CULTURES.

**auxins:** Plant growth regulators with properties similar to indole-acetic acid (IAA). Synthesized from tryptophan, IAA occurs in all plants; indolebutryric acid (IBA) occurs in maize and brassicas (Srivastava, 2002; Taiz and Zeigler, 2002). Synthetic auxins include 2,4-dichlorophenoxycetic acid and 2-methyl-3,6-dichlorobenzoic acid (Dicamba), which are used as herbicides (Basra, 2000) and in tissue culture (George, 1996; Gaba, 2005). IAA is synthesized in meristems, young leaves, and developing fruits and seeds and moves basipetally. Auxin is also synthesized in older leaves and moves apolarly in the phloem. Auxins promote growth in coleoptiles and stems and inhibit growth in roots. Auxin is involved in tropic responses, regulates apical dominance, delays leaf abscission, and promotes fruit development. Auxin induces xylem in vitro and in vivo. *See* ABSCISSION; PLANT GROWTH REGULATOR.

**auxin transport:** The polar movement of auxin from shoot apex to the root tip and back to the stem base (Taiz and Zeigler, 2002). *See* AUXIN INHIBITORS; AUXINS.

**auxotrophic growth:** The growth of an organism or plant tissue via fulfillment of a nutritional requirement for some compound(s) it cannot synthesize.

**AVG:** *See* AMINOETHOXYVINYLGLYCINE.

**axenic culture:** An uncontaminated or pure culture. This term is sometimes used to describe plant tissue cultures that are free of pathogens and visible contamination and have been tested for the presence of cultivable contaminants. Axenic status is in practice arguably impossible to confirm; at best cultures can be claimed to be free of cultivable contaminants and of microorganisms for which the tissues were tested (Cassells, 1997). *See* ASEPTIC CULTURES; CULTURE INDEXING; PATHOGEN INDEXING.

**axillary buds:** Buds in the junction with the stem of a leaf or similar organ that develop into stems. Axillary bud induction in vitro from excised meristems is a common strategy for micropropagation. *See* ADVENTITIOUS BUDS; AXILLARY SHOOTS; MICROPROPAGATION.

**axillary shoots:** Shoots that arise from axillary buds. *See* AXILLARY BUDS.

**B:** *See* BORON.

**B5 medium:** *See* GAMBORG'S B5 MEDIUM.

**BA:** *See* BENZYLADENINE.

*Bacillus:* A genus of gram-positive bacteria that is ubiquitous in nature (soil, water, and airborne dust). It is able to produce resistant endospores when environmental conditions are stressful. The presence of *Bacillus* is an indicator of failure of good working practices in a tissue culture laboratory. *See* GOOD LABORATORY PRACTICE.

*Bacillus thuringiensis:* Bt; an insecticidal bacterium marketed globally for control of plant pests—mainly caterpillars of the Lepidoptera order (butterflies and moths). Bt products represent about 1 percent of total pesticide sales. The commercial Bt products are powders containing a mixture of dried spores and toxin crystals. The crystal protein is highly insoluble and is safe for humans, higher animals, and most insects. It is solubilized in reducing conditions above ~pH 9.5. These conditions are commonly found in the midgut of lepidopteran larvae; consequently, Bt is a highly specific insecticidal agent. Bt genes have been used in plant transformation. *See* GENETIC ENGINEERING.

**bacitracin:** Antibiotic originally obtained from *Bacillus subtilis.* Caution: can cause eye and skin irritation, nausea, vomiting, and diarrhea (Walsh, 2003). *See* ANTIBIOTICS.

**backcross:** Breeding of a hybrid to one of its parents or to a genetically identical plant (Allard, 1999).

**bacterial elimination:** Bacteria may be eliminated by meristem culture, although it may be beneficial first to treat the infected plant with antibiotics to reduce apical contamination (Cassells et al., 1988). Where the apical explant is contaminated, it may be placed on

doi:10.1300/5648_02

antibiotic-containing medium (Barrett and Cassells, 1994). *See* ANTI-
BIOTICS; CONTAMINATION; PATHOGEN ELIMINATION.

**bacterial identification:** Following gram staining, plant-associated
bacteria are routinely identified by a combination of biochemical
tests and culture on selective media (Lelliot and Stead, 1987; Schaad
et al., 2001). Pathogenicity is confirmed by inoculation of indicator
plants. ELISA kits are available for many plant pathogenic bacteria;
alternatively, bacterial pathogens can be sent to service providers for
identification using DNA fingerprinting or fatty acid profiling (Stead
et al., 2000). Contamination arising from human-associated or environmental microorganisms can be identified with an appropriate API
kit (Cassells and Doyle, 2004). *See* API KITS; DNA FINGERPRINTING;
ENZYME-LINKED IMMUNOSORBENT ASSAY; FATTY ACID PROFILING;
GRAM STAIN.

**bacterial spores:** Resistant spores formed by some bacteria when
stressed (Goto, 1992). *See BACILLUS.*

**bactericide:** A substance that kills bacteria; a bactericidal agent. *See*
ANTIBIOTICS.

**bacteriological medium:** A medium for culturing bacteria, usually a
defined medium for specific bacterial species (Lelliot and Stead,
1987; Schaad et al., 2001).

**bacteriophage:** Viruses that infect bacteria (Kutter and Sulakvelidze,
2004).

**bacteriostat:** A chemical or other agent that blocks bacterial growth
and multiplication without killing the bacterium; a bacteriostatic
agent (Walsh, 2003). *See* ANTIBIOTICS.

**bacterium, pl. bacteria:** An organism whose cells contain neither a
membrane-bound nucleus nor other membrane-bound organelles such
as mitochondria and plastids. Bacteria include the Eubacteria ("true"
bacteria) and the Archaea. Characteristics of Eubacteria are the
absence of mitochondria or chloroplasts and a single chromosome

consisting of a closed circle of double-stranded DNA with no associated histones. If flagella are present, they are made of a single filament of the protein flagellin; the "9 + 2" tubulin-containing microtubules found in the eukaryotes are not present. The ribosomes differ in their structure from those of eukaryotes, a difference that is exploited by some antibiotics. Bacteria have a rigid cell wall made of peptidoglycan. Inhibition of wall synthesis is another site of antibiotic inhibition. The plasma membrane is a phospholipid bilayer but contains no cholesterol or other steroids. No mitosis takes place; reproduction is mostly asexual. Rather than meiosis, DNA may be transferred unidirectionally by conjugation. Many bacteria form spores when stressed that are very resistant to adverse conditions of dryness and temperature and may remain viable for long periods (Balows, 1992; Lengeler et al., 1999). *See* ARCHAEBACTERIA; PROKARYOTE.

**bacticinerator:** An electrical appliance generating high temperatures for sterilizing instruments. The device is used within laminar flow hoods to replacing the open flames of Bunsen burners, especially in the absence of a gas source. *See* GLASS BEAD STERILIZER; INSTRUMENT STERILIZATION.

**balance:** Instrument for weighing; to weigh, adjust, or equilibrate.

**ballast:** A step-up transformer increasing voltage and decreasing amperage to lighting systems but generating a great deal of heat and thus needing to be excluded from culture areas.

**BAP:** *See* BENZYLADENINE.

**basal:** At the base of a plant or organ. *See* BASAL MEDIUM.

**basal cell:** The larger vacuolated cell formed from the first division of the zygote that gives rise to the suspensor. *See* EMBRYOGENESIS; SUSPENSOR CELLS.

**basal medium:** A plant tissue culture medium without growth substrate (e.g., sucrose), growth regulators, or agar. It contains macronutrients, micronutrients, and organic components (George, 1993, 1996).

**basal transport:** Transport away from the apex in both roots and shoots.

**base:** Point on which an organism rests or its lowest part; a nitrogenous base in nucleic acid chains; a water-soluble, alkaline chemical compound that can release hydroxyl ions on dissociation and reacts with an acid to form a salt. A weak base forms a buffer with its conjugate acid (Metzler, 2001).

**base-pairing:** The pairing between complementary nitrogenous bases in nucleic acid chains: cytosine with guanine or adenine with thymidine/uracil (Russell, 2002). *See* DEOXYRIBONUCLEIC ACID; RIBOSE NUCLEIC ACID.

**basipetal:** Developing or flowering successively toward the base; describes the transport of substances and ions toward the base of the plant. *See* ACROPETAL.

**batch culture:** A cell suspension grown in a fixed volume of liquid medium and often showing five phases of growth following inoculation: lag, exponential growth, linear growth, deceleration, and stationary phases (George, 1993). Inocula of successive subcultures are of similar size and contain approximately the same cell mass at the end of each passage (Stepan-Sarkissian, 1991).

**bead culture:** Culture of protoplasts or cells in alginate beads (Scragg, 1991b). *See* CELL IMMOBILIZATION; PROTOPLAST CULTURE.

**bead sterilizer:** *See* GLASS BEAD STERILIZER.

**Benlate:** *See* BENOMYL.

**benomyl:** Benomyl is a systemic, broad-spectrum benzimidazole carbamate fungicide, also known by the trade names Benlate and

Tersan. It controls a wide range of diseases of fruits, nuts, vegetables, field crops, turf, and ornamentals. Powdery mildew, apple scab, and gray mold fungus are well controlled. Benomyl is effective against mites and can be used as a preharvest systemic fungicide and as a postharvest dip or dust treatment for the protection of fruits, seeds, and vegetables in storage. It breaks down to carbendazim and butyl isocyanate. The latter is toxic to tissues in culture and it is recommended that carbendazim be used or that the benomyl solutions be boiled to drive off the butyl isocyanate before addition to culture media (Hauptmann et al., 1985). *See* ANTIMYCOTICS; FUNGICIDES.

**benzalkonium chloride:** *See* ZEPHIRAN.

**benzothiadiazol:** Synthetic analogue of sialic acid providing good protection against downy mildew (Tosi et al., 1999).

**benzyladenine:** 6-Benzyladenine (BA) or 6-benzylaminopurine (BAP); a very active synthetic cytokinin used in plant tissue culture media to induce axillary bud or callus proliferation (Nogu, et al., 2003). Found naturally in some species, e.g., *Lycopersicon esculentum* (Gahan and Wyndaele, 2000).

**Bergman plating technique:** A method in which cells are mixed with warm agar just prior to setting in a petri dish (Bergman, 1960).

**beryllium:** Be; a highly toxic metal that appears to compete with magnesium at many enzymes sites (Marschner, 1994). As it is a small ion it is more likely to be present as a covalent complex (Metzler, 2001).

**β-galacturonidase gene:** A reporter gene used in genetic transformation. *See* GENETIC ENGINEERING.

**β-glucuronidase gene:** Syn. *GUS* gene; a reporter gene used in plant genetic engineering (Glick and Pasternak, 2003; Slater et al., 2003). *See* GENETIC ENGINEERING.

**β-particle:** Negatively charged electron emitted over a range of velocities during radioactive decay. Such particles emitted, e.g., from tritium are of value in autoradiography for localization of molecules in tissue sections (Gahan, 1984).

**biennial:** Describes a two-year life cycle in which plants grow and store food reserves during the first season and flower during the second season. The process requires cold temperatures in the first year. *See* VERNALIZATION.

**binding sites:** Sites of binding of substrates to enzymes; of action of plant bioregulators or hormones; of antibodies binding to their antigens; of organelles binding to motor molecules; and of macromolecules binding to macromolecules (Metzler, 2003). *See* ETHYLENE RECEPTORS.

**binucleate:** Having two nuclei per cell.

**bioassay:** A biological test performed on living cells or organisms, often to detect minute amounts of compounds that can affect or are necessary for growth. Used to screen compounds of pharmaceutical interest (Streibig and Kudsk, 1992). *See* ETHYLENE TRIPLE RESPONSE.

**biocide:** Disinfectant with fungicidal, insecticidal, and algaecidal activities. *See* SURFACE STERILANTS.

**biocontrol:** *See* BIOLOGICAL CONTROL.

**bioconversion:** *See* BIOTRANSFORMATION.

**biofilms:** Formed by the adhesion of bacteria to surfaces through the production of extracellular polysaccharides. Although many bacteria can grow in a free-living "planktonic" state, it is quite common for them to adhere to surfaces in biofilms. The biofilm may contain more than one species of microorganism. Microorganisms in biofilms may be resistant to the action of common surface sterilizing agents (Ghannoum and O'Toole, 2004). *See* SURFACE STERILIZATION.

**bioinformatics:** Originally, the application of mathematics and computer science in genomics research; now broadened to include proteomic and metabolomic research (Mount, 2002). *See* GENOMICS; METABOLOMICS; PROTEOMICS.

**biolistics:** The introduction of DNA into plant cells on metal particles that are forcibly discharged into the tissue (Gray, Compton, et al., 2005). *See* PARTICLE BOMBARDMENT.

**biological control:** The control of pathogens, pests, and weeds by biological means; usually refers to the use of bacterial, fungal, or viral inoculants and beneficial insects (predators, pests, and parasitoids) (Gurr and Wratten, 2002). Crop rotation, plant breeding, and genetic engineering are included in the broader definition, as are the inoculation of microplants with beneficial organisms in vitro or at establishment (van Driesche and Bellows, 1996; Mukerji and Upadhyay, 1999). *See* MYCORRHIZAL FUNGI; PLANT-GROWTH-PROMOTING RHIZOBACTERIA.

**biological nitrogen fixation:** *See* NITROGEN FIXATION.

**bioreactor:** Syn. fermenter; a device made of either glass or steel in which organisms are grown in a sterile environment. The temperature-controlled structure permits the sampling of the continuously mixed cells/medium. A bioreactor offers the ability to change the medium, an adequate air supply, and a means of monitoring the internal environment for factors such as pH and temperature. The many designs permit the culturing of a range of cell types (Scragg, 1991a; Figure 5). Bioreactors have been adapted for plant tissue culture and micropropagation (Hvoslef-Eide and Preil, 2004). *See* BIOTRANSFORMATION; TEMPORARY IMMERSION.

**biostatic:** *See* ANTIBIOTICS; BIOCIDE.

**biosynthesis:** Synthesis of molecules in a living organism, often through a series of linked steps in a biosynthetic pathway (Metzler, 2001).

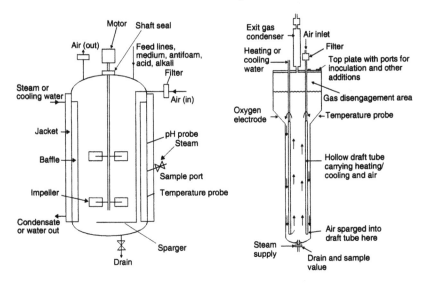

FIGURE 5. Examples of bioreactors: (left) a stirred bioreactor with aseptic air supplied via a sparger; (right) an air-lift bioreactor in which the aseptic air is used to maintain the plant cells or tissues in suspension. *Source:* Stafford and Warren (1991). *Plant Cell and Tissue Cultures.* Milton Keynes, UK: Open University Press, Figures 9.1 & 9.4. *See* PLANT BIOTECHNOLOGY.

**biotechnology:** The use of biological processes, often enhanced by genetic engineering in vivo and in vitro, in industrial and manufacturing processes, e.g., the production of alcohol by yeast fermentation, of secondary compounds by plant cell cultures, and of pharmaceutical compounds by whole plants. Biotechnology is seen by many as centered on advances in recombinant DNA technology (Glick and Pasternak, 2003; Slater et al., 2003). *See* PLANT BIOTECHNOLOGY.

**biotic factors:** Environmental factors such as water, temperature, and soil type that influence plant growth. *See* ABIOTIC FACTORS.

**biotic stress:** Stress caused by pathogens and pests characterized by the triggering of plant defenses involving the induction of phytoalexins and pathogenesis-related proteins (Orcutt and Nilsen, 2000; Taiz and Zeigler, 2002; Strange, 2003; Figure 1). *See* ABIOTIC STRESS; INDUCED RESISTANCE; PHYTOALEXINS.

**biotin:** Vitamin H; water-soluble coenzyme present in all cells and important in general metabolism; often added to plant tissue culture media. Factor promoting growth and respiration of *Rhizobium trifolii* in clover root nodules (Metzler, 2001).

**biotin labeling:** A nonradioactive DNA fluorescent probe based on the incorporation of biotinylated dUTP (deoxyuridine triphosphate) into DNA (Larsen et al., 1998).

**biotization:** Inoculation of aseptic plants in vitro or at establishment with beneficial microorganisms, e.g., mycorrhizal fungi or plant-growth-promoting rhizobacteria (Vestberg et al., 2002). *See* BIOLOGICAL CONTROL; MYCORRHIZAL FUNGI; PLANT-GROWTH-PROMOTING RHIZOBACTERIA.

**biotransformation:** The process by which the functional groups of organic compounds are modified by living organisms; used in the mixed biochemical/biological synthesis of pharmaceuticals where a low-value substrate is converted to a higher-value product or where stereospecificity is required in a compound (Stepan-Sarkissian, 1991). It is important that (1) the substance under investigation is not toxic to plant cells, (2) the rate of its production exceeds the rate of its further metabolism by the cells, (3) the substrate can readily enter the cells and the relevant metabolic compartment, and (4) the product can be readily extracted from the culture. Plant cells may be used in suspension together with the compound to be transformed, with the cells being collected at the end of the incubation; or the plant cells may be immobilized, with the required compound being removed from the medium; or the relevant enzyme can be isolated from the cells and used in an immobilized form. A broad range of compounds can be prepared, including phenylpropanoids, mevalonates, and alkaloids, and xenobiotics can be transformed or degraded by plant cells (Scragg, 1991a,b; Stepan-Sarkissian, 1991). *See* BIOREACTOR; CELL IMMOBILIZATION.

**bivalents:** A pair of homologous synapsed chromosomes during the first meiotic division (Russell, 2002).

**blackening:** Leaching of phenolic exudates into culture medium, which turn black/purple on oxidation, concomitant with culture necrosis and death.

**blade:** Expanded part (lamina) of a leaf or petal (Esau, 1977; Mauseth, 1998).

**bleach:** A disinfectant fluid or powder containing either calcium or sodium hypochlorite to remove color (bleach). Used to sterilize work surfaces, instruments, and plant materials for tissue culture. *See* SURFACE STERILANTS.

**bleached:** Describes a loss of chlorophyll's green color, often as a result of oxidative damage (Taiz and Zeigler, 2002).

**blue light response:** Light of wavelengths 425–490 nm suppresses hypocotyl growth and is involved in phototropism and stomatal and chloroplast movement (Taiz and Zeigler, 2002). CRY1 is a specific photoreceptor for blue as opposed to UV light; it is a cryptochrome, i.e., a photosensory receptor mediating light regulation of growth and development in plants. It has a chromophore-binding domain that has a similar structure to DNA photolyase and a carboxyl terminal extension that contains a DQXVP-acidic-STAES (DAS) domain conserved from mosses through ferns to angiosperms (Shalitin, 2003). CRY1 has FAD and possibly deazaflavin linked as chromophores (Briggs and Liscum, 1997). *See* FLAVINE ADENINE DINUCLEOTIDE.

**B*n*:** Backcross generations (B1, B2 etc.) (Allard, 1999).

**B-nine:** *See* AMINOCYCLOPROPANECARBOXYLIC ACID.

**bolting:** Elongation of the stem of a rosette plant usually associated with premature flowering (Taiz and Zeigler, 2002).

**boric acid:** Syns. boracic acid, orthoboric acid; an antiseptic.

**boron:** B; an essential micronutrient involved in the functioning of the plasma membrane. Boron deficiency is associated with enhanced IAA oxidase activity, decreased levels of diffusible IAA, increased calcium deficiency, and enhanced production of oxygen free radicals

(Marschner, 1994). It is usually included in plant tissue culture media as borax (George, 1993, 1996; Haensch, 1999).

**borosilicate glass:** Soda lime glass containing ~5 percent boric oxide. It is stronger and more heat resistant than soda lime glass.

**botany:** The scientific study of all aspects of plants, including all microorganisms because of their involvement with higher plants (Graham et al., 2003; Mauseth, 2003).

**bovine serum albumin:** BSA; a protein isolated from bovine serum that is used as a protein molecular weight standard in gel electrophoresis and to block nonspecific binding sites in ELISA and immunocytochemistry. *See* ENZYME-LINKED IMMUNOSORBENT ASSAY; GEL ELECTROPHORESIS.

**bract:** Leaflike structure subtending a flower or inflorescence (Esau, 1977; Mauseth, 1988).

**branch:** An axillary shoot or lateral root.

**brassinosteroids:** Steroid-related plant growth regulators, e.g., brassinolide (Yokoda, 1997; Gaba, 2005).

**break:** Act of breaking; side shoot.

**breeding:** Means of improving plant (and animal) varieties through genetic alteration (Allard, 1999). *See* GENETIC ENGINEERING; GENETIC MANIPULATION; MUTATION BREEDING.

**bridge:** Filter paper or other form of support used with liquid culture medium in plant tissue culture (George, 1993).

**broadleaf plant:** A dicotyledonous (dicot) angiosperm (Heywood, 1993). *See* MONOCOTYLEDONAE.

**broad-spectrum medium:** A medium that has been developed to support a wide range of species or cultivars, as opposed to a medium optimized for a single cultivar (Cassells and Morrish, 1986).

**bromocresol purple:** An indicator changing from yellow to purple in the pH range 5.2–6.8.

**5-bromodeoxyuridine:** BRdU, BrDU; a nucleic acid base analogue readily incorporated by growing cells into their nucleic acids, where it suppresses the uptake of deoxythymidine. Such wild-type cells are readily killed by light during the selection of auxotrophs. Replicated chromosomes containing BRdU help with the identification of chromatid exchanges (harlequin figures), and fluorochrome-tagged BRdU is used in cell-cycle studies (Ormerod, 2004).

**bromoxynil:** A nitrile herbicide (Basra, 2000).

**browning:** Discoloration as a result of phenol oxidation at freshly cut surfaces of explants. Browning at later stages of culture may indicate nutritional or pathogenic problems leading to necrosis. *See* BLACKENING.

**BSA:** *See* BOVINE SERUM ALBUMIN.

**Bt:** *See BACILLUS THURINGIENSIS.*

**Büchner filter:** Porcelain filter with a flat, perforated base used in vacuum filtration.

**bud:** Undeveloped (embryonic), unemerged leaf, flower, or stem usually enclosed by reduced/specialized leaves (bud scales) (Esau, 1977; Mauseth, 1988); start of (incipient) growth; vegetative outgrowth from a yeast.

**budding:** First signs of seasonal growth; grafting of a bud onto a rootstock; formation of buds by yeast.

**bud scale:** A modified leaf that protects a bud (Esau, 1977; Mauseth, 1988). *See* MERISTEM CULTURE.

**bud sport, bud mutation:** Bud containing a somatic mutation leading to the formation of an atypical branch, fruit, or flower. Such characters can be retained by vegetative propagation (Hartmann et al., 2001). *See* CHIMERA.

**buffer:** A combination of a conjugate acid and its base resulting in a solution that resists pH change over a fixed pH range (buffering capacity) on the addition of an acid or a base by either binding or releasing $H^+$ ions (Metzler, 2001).

**bulb:** An underground food-storage or reproductive organ with a short stem bearing specialized fleshy leaves, e.g., onion, tulip. Scale leaves may completely cover the fleshy leaves (tunicate), as in tulips, or may overlap each other and cover small portions of the fleshy leaves (scaly), as in lilies (Esau, 1977). An incandescent lamp. *See* ARTIFICIAL LIGHT.

**bulbil:** Bud formed in the position of a flower bud but containing no floral leaves; readily released from the parent plant to grow into a new plant. Commonly found in some lilies and onions (Esau, 1977).

**bulblet:** A bulb formed in vitro (Takayama et al., 1991). *See* MICROBULB.

**bundle sheath cells:** Layer(s) of cells surrounding the vascular bundles of stems and the small veins of leaves (Mauseth, 1988; Esau, 1997). *See* C4 PLANTS.

**bung:** Cork or rubber stopper.

**Bunsen burner:** Named for R. W. Bunsen 1811–1899; used for sterilizing tools and container openings using a hot flame from the ignition of natural gas and air. In the absence of a sterile hood, a sterile environment will occur between two lighted Bunsen burners placed 1 m apart. *See* GLASS BEAD STERILIZER.

**burette:** Graduated glass or plastic tube with a tap for dispensing measured volumes of liquid.

**burgeon:** Sprout, bud, grow, flourish.

**B vitamin:** *See* VITAMIN B COMPLEX.

**C3 plants:** Plants in which the first stable product of photosynthesis is the 3-carbon compound 3-phosphoglycerate (Taiz and Zeigler, 2002). *See* C4 PLANTS; CALVIN CYCLE.

**C4 plants:** Plants in which the first initial carboxylation is 4-carbon oxaloacetate. In these plants carbon fixation occurs in the mesophyll cells and reduction in the bundle sheath cells (Taiz and Zeigler, 2002). *See* C3 PLANTS.

**calcium:** Essential mineral involved in cell wall synthesis, membrane function, and cell signaling (Metzler, 2001). Taken up in the transpiration stream (Marschner, 1994). Low calcium uptake is associated with stomatal malfunction and hyperhydricity. *See* HYPERHYDRICITY; MINERAL NUTRITION; STOMA.

**calcium alginate:** A gel used to encapsulate cells, protoplasts, or somatic embryos. It is formed by dropping sodium alginate containing the cells, protoplasts, or embryos into a bath of calcium chloride, where it solidifies (Lindsey and Yeoman, 1983). It is used to form an artificial endosperm based on an appropriate culture medium or to encapsulate a somatic embryo in synseed (artificial seed) production (Redenbaugh, 1993). *See* ARTIFICIAL SEED.

**calcium hypochlorite:** An oxidizing agent widely used to sterilize equipment and plant material; usually used with a wetting agent or as a dilution of a commercially formulated household bleach (George, 1993). *See* SURFACE STERILANTS.

**calcofluor:** One of several fluorescent brightening agents, e.g., calcofluor white, used as a stain to detect wall reformation about protoplasts by reacting them with the newly formed cellulose (Gahan, 1984; Warren, 1991). *See* PROTOPLAST; SOMATIC HYBRIDIZATION.

**calibrated dichotomous sensitivity test:** CDS test; a method for determining the antibiotic susceptibility of microorganisms (Bell et al., 1999).

doi:10.1300/5648_03

**calliclones:** Cell colonies derived from single plated cells (Cassells et al., 1987).

**callose:** An insoluble substance found in plant cell walls, in stigmas on pollination by foreign pollen, and formed in response to cell injury. It consists of chains of $\beta$-1,3 linked glucose units (Strange, 2003).

**callus tissue:** A growth of nonspecialized plant cells (George, 1993). *See* ADVENTITIOUS REGENERATION.

**Calvin cycle:** The pathway from the reduction of carbon dioxide to ribulose-1,5-bisphosphate in photosynthesis. It involves the carboxylation of ribulose-1,5-bisphosphate by rubisco (**ribulose bis**phosphate **carboxylase/oxygenase**) (Metzler, 2003). *See* RIBULOSE BISPHOSPHATE CARBOXYLASE.

**CAM:** *See* CRASSULACEAN ACID METABOLISM.

**cambium:** Secondary meristem, a layer of meristematic cells between the phloem and the xylem that expands the girth of plants. A specialized cambium forms the cork layer (Esau, 1977; Mauseth, 1988). *See* APICAL MERISTEM; LATERAL MERISTEM.

**caramelization:** Nonenzymatic browning occurring at high temperatures and consisting of a succession of dehydration, condensation, and polymerization reactions involving sugars. Galactose, sucrose, and glucose caramelize at ~160°C, fructose at 110°C, and maltose at ~180°C. Caramelization occurs when plant tissue culture media are exposed to excessive temperature or pressure during sterilization.

**carbendazim:** The fungicide benomyl is converted to butyl isocyanate and carbendazim (MBC) in plant tissues; the latter is the active principle. Carbendazim binds to tubulin, interfering with the mitotic apparatus and inhibiting nuclear division (Agrios, 1997). *See* BENOMYL.

**carbenicillin:** A semisynthetic derivative of penicillin that inhibits susceptible strains of gram-negative bacteria, such as *Pseudomonas* (Walsh, 2003). *See* ANTIBIOTICS.

**carbohydrates:** Polymers of sugars (Dey and Harborne, 1996; Bryant et al., 1999; Metzler, 2001). *See* SUGAR.

**carbol fuchsin:** Chromosome stain comprising basic fuchsin, phenol, glacial acetic acid, and formalin (Darlington and La Cour, 1976).

**carbon dioxide:** A constituent of the atmosphere (350 ppm by volume) that may limit photosynthesis at high light intensities, requiring supplementation for optimal rates of photosynthesis (Hartmann et al., 2001). *See* AUTOTROPHIC CULTURE; CARBON DIOXIDE ENRICHMENT; PHOTOSYNTHESIS.

**carbon dioxide compensation point:** *See* COMPENSATION POINT.

**carbon dioxide enrichment:** Supplementation of carbon dioxide in greenhouses provided at high light intensities where ambient carbon dioxide is limiting for photosynthesis (Hartmann et al., 2001). *See* AUTOTROPHIC CULTURE.

**carbon growth source:** The substrate, usually sucrose, for plant growth in heterotrophic or mixotrophic culture (George, 1996; Beyl, 2005). *See* HETEROTROPHIC GROWTH; MIXOTROPHIC CULTURES.

**carotenoids:** Orange photosynthetic and light-stress-protecting compounds, polymers of alkenes (Wink, 1999; Taiz and Zeigler, 2002).

**carpel:** Structure that contains the plant ovules. *See* OVULE CULTURE.

**casamino acids:** Amino acids derived from casein. *See* CASEIN HYDROLYSATE.

**casein hydrolysate:** A hydrolysis product of casein, a milk protein, when broken down by enzymes or acids. As the composition is variable it is important in documentation to specify the method used to prepare the hydrolysate or the supplier's details. Casein hydrolysate is sometimes included in plant tissue culture media. *See* MENARD'S MEDIUM.

**caulogenesis:** Adventitious regeneration of shoots (George, 1996). *See* ADVENTITIOUS REGENERATION; ORGANOGENESIS; RHIZOGENESIS.

**C-banding:** *See* CENTROMERE BANDING STAIN; CHROMOSOME.

**CCC:** *See* CHLORMEQUAT.

**cDNA:** Copy DNA; DNA copies of an RNA template synthesized by reverse transcriptase (Russell, 2002).

**cDNA library:** The collection of cDNA copies of the mRNA of a cell (Russell, 2002).

**cefotaxime:** A third-generation cephalosporin active against gram-negative enteric bacteria (Walsh, 2003). Cefotaxime shows low toxicity to plant tissue cultures (Mihaljevic et al., 2001). *See* ANTIBIOTICS.

**cell commitment:** *See* DETERMINATION.

**cell culture:** The culture of isolated cells (Curry and Cassells, 1998).

**cell cycle:** The cycle of mitosis (nuclear division) and cytokinesis (cell division). Divided into G1, S, G2, and M periods, where S is the period of DNA synthesis (Russell, 2002). *See* MITOSIS.

**cell cycle arrest:** Cessation of cell division as found in noncycling cells and represented by a long G1 phase of the cell cycle, though cells may also opt out of the cycle into a quiescent state represented by G0 or Gq (Russell, 2002). *See* MITOSIS.

**cell death:** *See* APOPTOSIS.

**cell division:** *See* CELL CYCLE; MITOSIS.

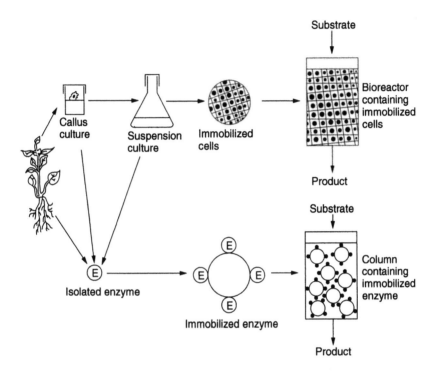

FIGURE 6. Cell and enzyme immobilization (Stepan-Sarkissian, 1991). *See* CELL IMMOBILIZATION.

**cell immobilization:** The process of confining catalytically active cells in a reactor system, preventing their entry into the mobile phase, which carries the substrate and product. The cells are commonly immobilized by adsorption to, or entrapment in, calcium alginate beads (Scragg, 1991b; Figure 6). *See* CALCIUM ALGINATE.

**cell suspension culture:** The culture of cells in liquid media. The cells are usually present as clumps. Aeration is achieved by shaking the culture vessel or by forced aeration. Cell growth passes through a stationary phase and an exponential growth phase to a lag phase (batch culture) or, if the medium is continuously renewed, a phase of continued growth (continuous culture). Cell suspension cultures are used for somatic embryogenesis and production of secondary metabolites (Scragg, 1991a). *See* BIOREACTOR.

**cellulases:** Enzymes that hydrolyze cellulose. Endocellulases hydrolyze glucose chains internally; terminal (extra)cellulases hydrolyze chains from the ends (Strange, 2003).

**cellulose:** A polymer composed of glucose residues linked by β-1,4-glycosidic bonds. It is the major component of plant cell walls (Brett and Waldron, 1989; Carpita et al., 2001). *See* PROTOPLAST ISOLATION.

**cell wall:** *See* PLANT CELL WALL.

**centromere:** A heterochromatin-associated region of the chromosome that bridges the chromatids prior to separation at anaphase of mitosis or anaphase 2 of meiosis. Centromeres are used in the identification of chromosomes for karyotyping (Dyer, 1979; Fukui and Nakayama, 1996). *See* HETEROCHROMATIN; KARYOTYPE ANALYSIS; MEIOSIS; MITOSIS.

**centromere banding stain:** C-banding; a selective chromosome banding staining technique involving Giemsa staining of the DNA after denaturation or extraction by treatment with alkali, acid, salt, or heat. Only heterochromatic regions close to the centromeres and rich in satellite DNA stain (Fukui and Nakayama, 1996). *See* GIEMSA.

**2-CEPA:** *See* 2-CHLOROETHYLPHOSPHONIC ACID; ETHEPHON.

**cephalosporins:** A group of more than 20 antibiotics derived from *Cephalosporium* spp. related to penicillin. Cephalosporins inhibit gram-positive and gram-negative bacteria by inhibiting bacterial cell wall synthesis. Overuse of cephalosporins has led to bacterial resistance to the drugs (Walsh, 2003). *See* ANTIBIOTICS.

**certification schemes:** *See* PLANT HEALTH CERTIFICATION.

**cetrimide:** Syns. cetrimonium bromide, CTAB, CTABr, HTAB; a cationic detergent that is a powerful disinfectant. *See* SURFACE STERILANTS.

**CFU:** *See* COLONY-FORMING UNIT.

**character:** A sexually inherited trait (Allard, 1999).

**charcoal:** The carbonaceous residue from the combustion of wood or animal matter with the exclusion of air. *See* ACTIVATED CHARCOAL.

**chelated iron:** Iron bound to a chelating agent such as ethylenediaminetetraacetic acid (EDTA). Chelation is used to maintain iron in a soluble form (George, 1993, 1996; Marschner, 1994). *See* ETHYLENEDIAMINETETRAACETIC ACID.

**chelating agents:** Organic compounds capable of forming a complex with cations through two or more atoms of the organic compound. The most important commercial family of chelating agents is ethylenediaminetetraacetic acid (EDTA) and its various sodium salts, made from ethylenediamine (EDA); also organic acids, e.g., citric, malic, and oxalic, via their $COO^-$ groups.

**chemical fusion:** The fusion of protoplasts using a chemical fusogen such as polyethylene glycol. *See* ELECTROFUSION; POLYETHYLENE GLYCOL; PROTOPLAST FUSION.

**chemically defined medium:** A medium in which all the components are chemically characterized as opposed to media containing uncharacterized materials such as coconut milk (George, 1993, 1996). *See* COCONUT MILK.

**chemical mutagenesis:** The induction of mutation by chemical mutagens such as ethylmethylsulfonate (EMS), in contrast to induction of mutation by physical mutagenesis, e.g., exposure to gamma rays (Micke and Donini, 1993; van Harten, 1998; Russell, 2002). *See* PHYSICAL MUTAGEN; SOMACLONAL VARIATION.

**chemical mutagens:** Chemicals that induce mutations (van Harten, 1998). Three types exist: (1) base analogues, (2) base-modifying agents, and (3) intercalating agents. Base analogues are similar to the bases found in DNA and may exist in tautomeric forms.

An example is 5-bromouracil, which has a 5-bromo in place of the 5-methyl in thymine and which in its normal state resembles thymine and pairs only with adenine but in its rare state pairs only with guanine, in the latter case resulting in a T–A to C–G base pair change. Base-modifying agents include nitrous acid, hydroxylamine, and methylmethanesulfonate (MMS). Nitrous acid removes the amino groups from guanine, cytosine, and adenine, which results in xanthosine, uracil, and hypoxanthine, respectively. Xanthosine has the same pairing behavior as guanine, but uracil pairs with adenine to give a C–G to T–A change, and hypoxanthine pairs with cytosine rather than uracil to give an A–T to G–C transition. Hydroxylamine hydroxylates cytosine specifically, which alters its pairing from adenine to guanine (C–G to T–A transition). MMS is representative of alkylating agents that add $-CH_3$ or $-CH_2CH_3$ at a number of locations, but mainly to the 6-oxygen in guanine to produce, e.g., $O^6$-methyl guanine, which pairs with thymine instead of cytosine, giving a G–C to A–T change. Intercalating agents, e.g., ethidium bromide, insert themselves (intercalate) between adjacent bases on one or both DNA strands; they pair with one of the DNA bases at random and the latter is paired at replication, e.g., if an A was inserted at random during replication, this would be paired with T and a base pair AT would be inserted into the sequence. Alternatively, the intercalating agent and the pairing base can be lost at replication to give a base deletion. The result of intercalation in either case is a frameshift mutation (Russell, 2002).

**chemotherapy:** The use of chemicals to control disease. The chemicals used include antibiotics. In plant chemotherapy these may be applied to the stock plant or incorporated into the tissue culture medium (Cassells, 1983, 1987; Barrett and Cassells, 1994). *See* ANTIBIOTICS; ANTIMYCOTICS; ANTIVIRAL COMPOUNDS; MERISTEM CULTURE; PATHOGEN ELIMINATION; THERMOTHERAPY; VIRUS ELIMINATION.

**chiasma, pl. chiasmata:** A cross-shaped configuration of the chromosomes at crossing over during the diplotene stage of meiosis (Dyer, 1979). *See* MEIOSIS.

**chilling requirement:** *See* VERNALIZATION.

**chimera:** A plant that consists of two or more genetically different types of cells (Tilney-Bassett, 1991). Conventionally, this refers to tissues arising from a chimeral apical meristem, usually expressed as variegation in the leaf or petal pigmentation. However, mutant sectors may also arise other than from apical division and are proposed as sources of variation expressed in adventitious shoots from such aberrant cells (D'Amato, 1977; George, 1993; Figures 7 and 8).

**chitin:** A polymer containing $N$-acetylglucosamine units found in the wall of members of the Eumycota (Strange, 2003). *See* FUNGUS.

**chlDNA:** *See* CHLOROPLAST DNA.

**chloramine:** Syns. chloramine T, chlorasan, clorosan; one of several compounds containing chlorine and nitrogen used as an antiseptic in wounds. Chloramine is a toxic gas created by the combination of ammonia and sodium hypochlorite bleach. Chlorine is being phased out in favor of chloramine for nonpotable water sterilization. *See* SURFACE STERILANTS.

**chloramphenicol:** An antibiotic originally isolated from *Streptomyces venezuelae;* it is now chemically synthesized. Chloramphenicol binds reversibly with the large ribosomal subunit of bacteria and eukaryotes. The bacterial and eukaryotic mitochondrial ribosomes are more sensitive than the eukaryotic cytosolic ribosomes, but host protein synthesis is also decreased. It binds to the peptidyl transferase enzyme to inhibit transfer of the growing polypeptide to the next amino acid occupying the acceptor site (Walsh, 2003). *See* ANTIBIOTICS.

**chlorine:** Cl; essential mineral; a component of chlorophyll (Marschner, 1994). *See* PLANT TISSUE CULTURE MEDIA.

**chlormequat:** CCC; a plant growth retardant (Basra, 2000). Used as a pretreatment for stock plants to enhance explant responses in vitro, to prolong culture life, and to promote growth in vitro (George, 1993).

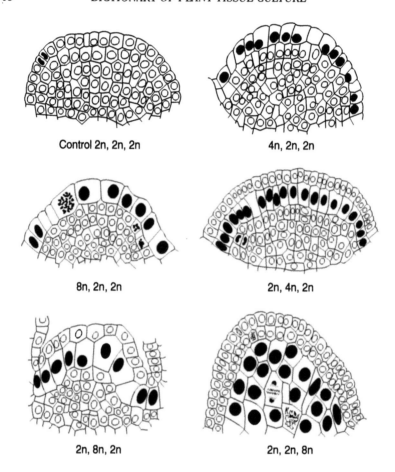

Control 2n, 2n, 2n          4n, 2n, 2n

8n, 2n, 2n          2n, 4n, 2n

2n, 8n, 2n          2n, 2n, 8n

FIGURE 7. Experimental evidence for the Tunica-Corpus hypothesis that the primary aboveground tissue originates from layered initial cells at the apex. Mainly periclinal division in the outer layer (Tunica L1) gives rise to the epidermal layer, whereas periclinal and anticlinal division in the Tunica L2 and L3 give rise to the subepidermal and core tissues, respectively. Satina et al. (1940) produced a series of *Datura* periclinal chimeras, illustrated here, in which initial cells in the respective apical initial layer were polyploidized and the progeny cells mapped (see also Figure 8). *See* CHIMERA; HISTOGEN THEORY.

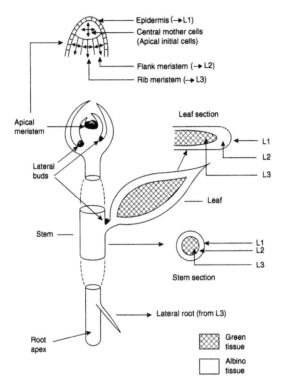

Epidermis (→L1)
Central mother cells (Apical initial cells)
Flank meristem (→ L2)
Rib meristem (→ L3)

Apical meristem

Leaf section

L1
L2
L3

Lateral buds

Leaf

Stem

L1
L2
L3

Stem section

Lateral root (from L3)

Root apex

Green tissue

Albino tissue

FIGURE 8. The Tunica-Corpus model of plant tissue development showing that the layered structure of the primary tissues arises from initials cells of the tunica and corpus (see also Figure 7). Mutation in the apical initials of the respective layers may give rise to a mutant cell lineage in the corresponding tissues, that is, to chimera formation. Here, the consequences of albino mutation in the initial cells of the L2 ("flank meristem") are illustrated in the tissue distribution of the progeny cells. The germ cells arise from the L2, consequently chimeras are not perpetuated via seed (Tilney-Bassett, 1991). Chimeras may be propagated via meristem culture but not stably via adventitious regeneration, which occurs from individual cells usually from one somatic layer (Cassells, 1992). In the example illustrated, adventitious shoots from the L2 would be without chlorophyll (albino). See also cover illustration. *See* CHIMERA; DIPLONTIC SELECTION.

**2-chloroethylphosphonic acid:** 2-CEPA; an ethylene-releasing plant growth regulator used as a defoliant (Basra, 2000). *See* ethylene.

**4-chlorophenoxyacetic acid:** A herbicide related to auxin (Basra, 2000; Taiz and Zeigler, 2002). *See* AUXINS.

**chlorophyll:** The main photosynthetic pigments in plants (Taiz and Zeigler, 2002). Chlorophyll absorbs light in the red and blue-violet regions (Figure 9). Chlorophyll is a tetrapyrrole chelate of magnesium. As most of a plant's nitrogen is in its chlorophyll, chlorophyll measurement is used as a parameter of nitrogen fertilizer requirement (Goel et al., 2003). Chlorophyll content is also used as a quality parameter for microplants.

**chlorophyll fluorescence:** When chlorophyll is excited by a photon, it passes the energy either to another molecule en route to photosynthesis or by releasing the energy, partly as heat and partly as the emission of a photon of lower energy, i.e., at a longer wavelength (Ridge, 2002). The characteristics of chlorophyll fluorescence are used to monitor photosynthetic efficiency and to characterize plant stress and have been used to evaluate the quality of microplants (Franck et al., 2001).

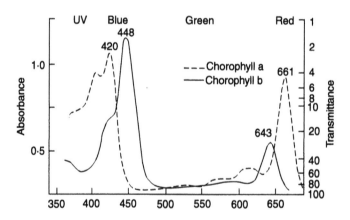

FIGURE 9. The light absorption spectrum of chlorophyll a and b (Bryce and Hill, 1993). Reprinted by permission. Natural light is of equal intensity at all wavelengths in the visible region of the electromagnetic spectrum; however, artificial light may be deficient in the blue and red regions. *See* ARTIFICIAL LIGHT; PAR LIGHT.

**chloroplast:** A DNA-containing green plastid containing the pigments involved in photosynthesis (Taiz and Zeigler, 2002; Mauseth, 2003). *See* ENDOSYMBIONT HYPOTHESIS; PHOTOSYNTHESIS.

**chloroplast DNA:** chlDNA; self-replicating small circular DNA strands with no associated protein, coding for a number of proteins associated with the chloroplast (Alberts et al., 2002; Buchanan et al., 2002). *See* ENDOSYMBIONT HYPOTHESIS.

**chlorosis:** The yellowing of older leaves associated with prolonged nitrogen deficiency.

**chromatid:** One of the two substructures of the chromosome visible in early prophase and metaphase of mitosis (Dyer, 1979). *See* CHROMOSOME.

**chromatin:** The complex of histone proteins with DNA, which forms visible chromosomes at mitosis (Turner, 2001). *See* EUCHROMATIN; HETEROCHROMATIN.

**chromatography:** The separation of molecules by passing a solvent across a stationary phase such as sheets of paper, silica gel plates, or columns packed with, e.g., hydroxyapatite or silica gel (Metzler, 2001). *See* PARTITION CHROMATOGRAPHY.

**chromosomal aberrations:** *See* CHROMOSOME MUTATIONS.

**chromosome:** Complex of DNA, RNA, proteins, and phospholipids consisting of one (pre-S phase) chromosome or 2 (post-S phase) chromatids with a centromere, two telomeres, and sometimes a secondary constriction (Dyer, 1979). The circular genomes of the chloroplast and mitrochondrion are also termed chromosomes; bacteria may also contain circular extranuclear DNA called plasmids (Russell, 2002). *See* CENTROMERE; NUCLEOLUS.

**chromosome mutations:** The main chromosomal mutations involve haploidy, aneuploidy, polyploidy, and structural rearrangements (Dyer, 1979; Fukui and Nakayama, 1996; Figure 10). If these

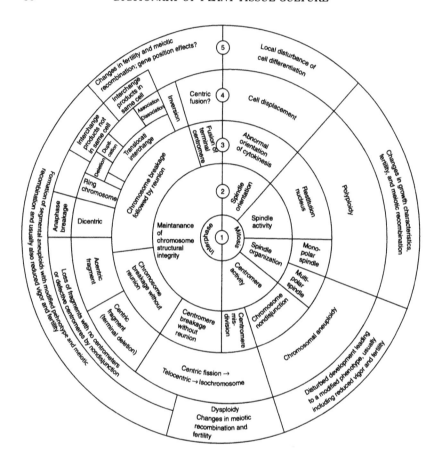

FIGURE 10. Chromosomal aberration—their origin and expression (Dyer, 1979). Reprinted by permission. These aberrations occur at low frequency in nature, are induced by physical and chemical mutagens and occur spontaneously at high frequency *in vitro* in some genotypes (Mohan Jain et al., 1998). *See* MUTA-TION BREEDING; SOMACLONAL VARIATION.

mutations occur in the apical initial cells and the mutant cell lineage is as fit as or fitter than the nonmutated cells, the mutation may be expressed via diplontic drift as chimeras (Balkema, 1972). If the mutation occurs outside the apex, the mutant cell lineage may not be expressed unless the mutant cells participate in cell division, adventitious regeneration in vitro, or both (D'Amato, 1977; George, 1993). *See* APICAL MERISTEM; CHIMERA; MUTATION BREEDING; SOMACLONAL VARIATION.

**circadian rhythm:** Biological cycles of high and low activity that show a regular daily cycle (Taiz and Zeigler, 2002; Mauseth, 2003).

**citric acid:** An organic carboxylic acid (Metzler, 2001).

**citric acid cycle:** Syns. Krebs cycle, TCA cycle, tricarboxylic acid cycle; a cycle of reactions that catalyze the oxidation of acetate to carbon dioxide and water and cellular energy in the form of $FADH_2$ and $NADH_2$. This energy may be passed to adenosine triphosphate (ATP) via the electron transport pathway (Metzler, 2001). *See* RESPIRATION.

**clonal propagation:** Propagation of plants true to type. Traditionally this was achieved by vegetative propagation in vivo (Hartmann et al., 2001). Micropropagation is an alternative in vitro cloning technique (Debergh and Zimmerman, 1991). *See* MICROPROPAGATION; VEGE-TATIVE PROPAGATION.

**clonal selection:** The selection of clones of lymphocytes in the immune response to obtain monoclonal antibodies (Roitt et al., 2001); can also refer to the selection and multiplication of hybrid, mutant, or transformed genotypes (Allard, 1999). *See* MONOCLONAL ANTIBODY.

**clone:** A population of cells or plants with identical genotypes. *See* STRAIN.

**cloning:** The multiplication of clones. In micropropagation genetic stability may be dependent on the in vitro cloning strategy used (George, 1993; Mohan Jain et al., 1998; Jayasankar, 2005). *See* SOMACLONAL VARIATION.

**cobalt:** Co; essential micronutrient, component of some enzymes (Marschner, 1994). *See* MINERAL NUTRITION.

**coconut milk:** The liquid component of the coconut endosperm. The composition of the milk changes during development. Coconut milk has high levels of amino acids, sugars, and cytokinins and contains mineral salts. Coconut milk was commonly added to tissue culture media to provide growth regulators and other compounds, but now defined components are preferred. The composition of coconut milk has been published (Tulecke et al., 1961). *See* PLANT TISSUE CULTURE MEDIA.

**coculture:** The culture of two organisms in the same culture vessel. This can involve the cocultivation of mutant cell lines, callus, or mutant shoots. *See* DUAL CULTURE; NURSE CULTURE.

**coding sequence:** The bases in an mRNA that are involved in specifying the amino acid sequence in a polypeptide chain during translation (Russell, 2002). *See* PROTEIN SYNTHESIS.

**codon:** A group of three adjacent nucleotides in a DNA or RNA sequence that specifies an amino acid in a polypeptide chain or acts as a termination sequence (Russell, 2002).

**colchicine:** A water-soluble alkaloid obtained from *Colchicum autumnale*. Colchicine inhibits microtubule assembly by binding to tubulin monomers to block spindle formation in mitosis and meiosis; in meiotic cells, this results in the formation of diploid gametes. It is used to produce polyploid cells in vivo and in vitro (Taji et al., 2001).

**cold storage:** The storage of biological material at below ambient temperature. Usually plant material is stored at around $+8°C$ (Towill, 2005). *See* GERMPLASM CONSERVATION.

**coleoptile:** A modified leaf that protects the young leaves of cereal seedlings as they grow through the soil (Esau, 1977; Mauseth, 1988).

**colony-forming unit:** CFU; the common origin for the cells of any colony (Prescott et al., 2001). A measure of viable spore count.

**commercial micropropagation:** The production for profit of plants by tissue culture (George, 1996; Suttle, 2005). *See* GOOD LABORATORY PRACTICE; MICROPLANT QUALITY; MICROPROPAGATION LABORATORY; PLANT HEALTH CERTIFICATION.

**commitment:** *See* DETERMINATION.

**compatible solutes:** Metabolites that accumulate in abiotic stress responses to adjust osmotic balance such as glycine betaine, mannitol, proline, and sorbitol (Basra and Basra, 1997; Lerner, 1999). *See* ABIOTIC STRESS.

**compensation point:** The concentration of carbon dioxide at which the carbon dioxide fixed in photosynthesis equals that produced by respiration (Ridge, 2002). *See* PHOTOSYNTHESIS.

**competence:** The ability of a cell or cells to respond to developmental signals in the predicted way (George, 1993; Schwarz et al., 2005; Figure 11). *See* EXPRESSION MEDIUM; INDUCTION MEDIUM.

**competitive exclusion:** The inability of a species to grow in a part of its habitat because of competition from a better-adapted organism (Tate, 2000). Inoculation of aseptic microplants or of microplants at establishment aims to exploit this phenomenon to exclude plant pathogenic fungi (Cassells, 2001; Vestberg et al., 2002). *See* BIOTIZATION.

**complementary base pairs:** The hydrogen bonding of adenine with thymine (in DNA), adenine with uracil (in RNA), and guanine with cytosine (in DNA and RNA) (Russell, 2002).

**complementation selection:** In protoplast fusion, the fusion of mutants with different defects, usually chlorophyll mutants, or antibiotic sensitivities, where only the hybrids survive, or grow faster, under selection (Warren, 1991; Veilleux, 2005). *See* PROTOPLAST FUSION.

**complex explants:** Explants containing different cell and tissue types (George, 1993). *See* THIN CELL LAYERS.

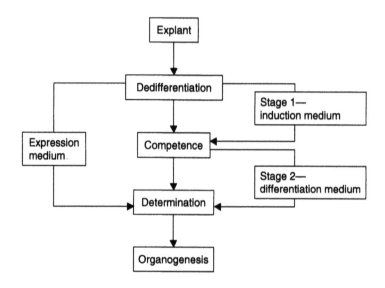

FIGURE 11. The stages in organogenesis. Explants from some genotypes undergo all stages on a single medium, in other cases, two distinct media are required (Schwarz et al., 2005). *See* DEDIFFERENTIATION; DETERMINATION; DIFFERENTIATION MEDIUM; EXPRESSION MEDIUM; INDUCTION MEDIUM; ORGANOGENESIS.

**compost:** Completely decayed organic matter. It is dark, odorless, and rich in nutrients. Compost can be used to amend soil for ornamental plants, trees, and potted plants (Hartmann et al., 2001).

**conditioned medium:** A medium that has been modified by the diffusion of metabolites from living cells (Warren, 1991). *See* NURSE CULTURE.

**confocal microscopy:** Method in which a series of confocal laser-scanned images of biological material in successive planes can be reconstructed by image analysis software to give 3D images (Sheppard et al., 1997; Murphy, 2001; Kaufman et al., 2004).

**conjugation:** The unidirectional transfer of DNA between bacteria (Prestcott et al., 2001; Russell, 2002).

**conservation:** *See* CRYOPRESERVATION; GERMPLASM CONSERVATION.

**constitutive gene:** A gene that is constantly expressed in a growing cell (Russell, 2002). *See* INDUCED GENE.

**contamination:** Pollution or infection. In plant tissue culture the term usually refers to contamination by microorganisms or microarthropods (Cassells, 1997, 2000a; Pype et al., 1997).

**continuous culture:** *See* CELL SUSPENSION CULTURE.

**cool-white:** A type of fluorescent tube used in plant growth rooms (George, 1993). *See* ARTIFICIAL LIGHT.

**Coomassie blue:** A stain used to detect and assay proteins. *See* GEL ELECTROPHORESIS.

**copper:** Cu; an essential micronutrient involved in electron transfer reactions and as a cofactor (Marschner, 1994). *See* MINERAL NUTRITION.

**corm:** A modified fleshy subterranean stem with paper-thin leaves, e.g., gladiolus (Esau, 1997; Mauseth, 2003).

**corpus:** Region of apical meristem giving rise to the L3 (See Figures 7 and 8). *See* APICAL MERISTEM; TUNICA.

**cotyledon:** The first leaf that develops from the plant embryo. Monocots have one, dicots two, and conifers several cotyledons. Cotyledons are involved in storage in dicots and gymnosperms and in nutrient transfer from the endosperm in most monocots (Mauseth, 1988; Esau, 1997). Frequently they exhibit greater morphogenic potential than leaves or other plant tissues (George, 1993). *See* SOMATIC EMBRYOGENESIS.

**crassulacean acid metabolism:** CAM; process by which plants fix carbon dioxide in the dark into 4-carbon malate, which is stored in the vacuole. In the light, malate releases carbon dioxide, which enters the Calvin cycle (Taiz and Zeigler, 2002).

**critical control points:** *See* HAZARD ANALYSIS CRITICAL CONTROL POINTS.

**crop gene pool:** The genes and their alleles present in a breeding population. The gene pool is subdivided into four components: GP1, the crop varieties and their wild ancestor; GP2, species that can be crossed with GP1 but with a low rate of success; GP3, species that can be crossed only with difficulty with GP1; and GP4, species with genes of interest that cannot be crossed with GP1 (Allard, 1999; Chrispeels and Sadava, 2003). *See* EMBRYO RESCUE; GENETIC ENGINEERING; IN VITRO POLLINATION; PROTOPLAST FUSION.

**cross-protection:** Originally, the suppression of the symptoms of a second virus following challenge inoculation of a virus-infected plant (Hull, 2002). Subsequently, the phenomenon was found to apply to the protective effects of a necrotizing pathogen against other pathogens (SAR, systemic acquired resistance) and of plant-growth-promoting rhizobacteria (PGPR) against pathogens (ISR, induced systemic resistance) (Strange, 2003). *See* BIOTIZATION; STRESS CROSS-TOLERANCE.

**crown gall:** Disease named after the large tumor-like swellings (galls) that typically occur at the crown of the plant, just above soil level, following infection (Sigee, 1993; Agrios, 1997). *Agrobacterium tumefaciens* causes crown gall disease of a wide range of dicots, especially members of the rosaceae. *See AGROBACTERIUM TUMEFACIENS.*

**cryopreservation:** Freezing of tissue and cells at −196°C in liquid nitrogen. The basic principles of cryopreservation apply to all cell types. Prior to freezing, the cells are treated with a cryoprotectant that protects the cells and their membranes from damage during the freezing process. After the cells have been immersed in the freezing medium containing the cryoprotectant, they are cooled very slowly prior to plunging them into liquid nitrogen. When required the cultures are thawed rapidly, the cryoprotectant is removed, and the cultures are cultured on conventional medium. Cryopreservation is increasingly being used to store germplasm of vegetatively propagated plants (Towill, 2005). *See* CRYOPROTECTANT.

**cryoprotectant:** A compound added to tissue culture media to prevent ice formation during cryopreservation. Cryoprotectants include ethylene glycol, dimethyl sulfoxide (DMSO), and glycerol (Towill, 2005). *See* CRYOPRESERVATION.

**cryptic contaminants:** Hidden contaminants that are not visible to the eye but may be indicated by halos around the explant. They are usually endophytic contaminants of the tissues that are inhibited by the strength or components of the medium. They may be expressed when the medium is diluted or the composition altered (Leifert and Cassells, 2001). *See* CULTURE INDEXING; GOOD LABORATORY PRACTICE.

**cultivable contaminants:** Contaminants that grow on plant tissue culture and common microbial media. These are generally plant-surface-associated microorganisms or common human-associated or general environmental microorganisms (Cassells and Doyle, 2004). *See* CULTURE INDEXING; FASTIDIOUS CONTAMINANTS.

**culture containers:** *See* TISSUE CULTURE VESSELS.

**culture indexing:** Indexing of tissue culture for microbial contaminants. It was traditionally based on microscopic examination with the use of staining techniques; now, biochemical test kits, fatty acid analysis, and DNA probes are used (Stead et al., 2000; Schaad et al., 2001). Contaminants usually originate from plant tissues or are laboratory environmental microorganisms. They can indicate faulty equipment or practices (Leifert and Cassells, 2001; Cassells and Doyle, 2004). *See* API KITS; BACTERIAL IDENTIFICATION; FATTY ACID PROFILING; FUNGAL IDENTIFICATION; GENETIC FINGERPRINTING; YEAST IDENTIFICATION.

**culture medium:** A medium for culturing microorganisms or plant cells, tissues, and organs. *See* BACTERIOLOGICAL MEDIUM; PLANT TISSUE CULTURE MEDIA.

**culture vessels:** Vessel in which aseptic culture is carried out. *See* BIOREACTOR; TISSUE CULTURE VESSELS.

**cutan:** A polymer made from long-chain hydrocarbons. A component of the cuticle (Carpita et al., 2001; Strange, 2003).

**cuticle:** A layer of cutin, cutan, and waxes that coats the outer surface of the epidermis and restricts the movement of water and gases in and out of the plant (Carpita et al., 2001; Strange, 2003). *See* MICROPLANT ESTABLISHMENT; MICROPLANT QUALITY.

**cutin:** A rigid polymer composed of hydroxylated fatty acids. The main component of the cuticle (Carpita et al., 2001; Strange, 2003).

**cuttings:** Pieces of stems or leaves used to vegetatively propagate plants (Hartmann et al., 2001). *See* MICROPROPAGATION; VEGETATIVE PROPAGATION.

**cybrids:** Cytoplasmic hybrids; partial hybrids produced in vitro where only the cytoplasm of the donor protoplast is sought to be retained in the acceptor protoplast. *See* CYTOPLASMIC INHERITANCE; PROTOPLAST FUSION.

**cycloheximide:** A protein synthesis inhibitor derived from *Streptomyces griseus.* Cycloheximide acts on 80S ribosomes to inhibit protein synthesis in the cytoplasm of eukaryotic cells but not in bacteria, mitochondria, or chloroplasts (Walsh, 2003). *See* ANTIBIOTICS; ENDOSYMBIONT HYPOTHESIS.

**cyclophysis:** differential effects exhibited by parts of a plant of the same age, but different positions on the plant which have been subjected to different physical exposure.

**cytokinin inhibitors:** Some chemical analogues of RNA bases antagonize the activity of cytokinins in certain physiological processes, but these have been little investigated in in vitro culture (George, 1993).

**cytokinins:** Natural cytokinins are plant growth regulators derived from the purine base adenine. Phenylureas, e.g., thidiazuron, also show cytokinin activity. Cytokinins are widely involved in plant development activities, usually with other plant hormones, e.g., auxin or ethylene (George, 1993). Among these activities are mitosis, differentiation of the shoot meristem, differentiation of the tissues of the root, leaf formation, chloroplast development, and leaf senescence (Taiz and Zeigler, 2002). *See* PLANT GROWTH REGULATOR.

**cytoplasm:** Region of the cell between the nuclear membrane and the plasmalemma containing membrane-bound organelles, ribosomes, cytoskeleton, and cytosol (Albert et al., 2002; Graham et al., 2003; Mauseth, 2003).

**cytoplasmic DNA:** Circular DNA in the mitochondrion and chloroplast (Russell, 2002). *See* CYTOPLASMIC INHERITANCE; ENDOSYMBIONT HYPOTHESIS.

**cytoplasmic inheritance:** Syn. maternal inheritance; the inheritance of traits coded for on cytoplasmic (chloroplast, mitochondrion) chromosomes. These are normally transmitted though the female gamete as the male gamete usually contains only the nuclear genes (Allard, 1999; Russell, 2002). Cytoplasmic genes can also be transferred by protoplast fusion (Veilleux et al., 2005). *See* CYBRIDS; PROTOPLAST FUSION.

**cytoplasmic male sterility:** A defect in mitochondrial DNA (mtDNA) that results in the formation of nonviable pollen, used in the production of $F_1$ hybrids (Allard, 1999).

**cytoplasmic organelles:** Membrane-enclosed structures in the cytoplasm, including mitochondria, chloroplasts, peroxisomes, Golgi bodies, vacuoles, endoplasmic reticulum, and transport vesicles (Graham et al., 2003; Mauseth, 2003).

**cytosine methylation:** *See* DNA METHYLATION.

**cytosol:** The aqueous colloidal phase of the cell (Graham et al., 2003; Mauseth, 2003). *See* CYTOPLASM.

**2,4-D:** *See* 2,4-DICHLOROPHENOXYACETIC ACID.

**damping-off:** Seed may rot either prior to emergence of the plumule or as a result of fungal infection by, e.g., *Pythium, Fusarium, Phytophthora* in cold and crowded conditions (Agrios, 1997; Strange, 2003). Microplants are susceptible to damping-off pathogens (Williamson et al., 1997). *See* BIOLOGICAL CONTROL; BIOTIZATION.

**dark ground illumination:** Limitation of light passing through the condenser of a microscope such that objects reflect light and are seen as light against a dark background.

**daylength:** An important parameter for plant growth in vivo, regulating ontogeny (Hartmann et al., 2001). Daylength in growth rooms is regulated by timers; a long day is usually 16 hours of light, a short day 8 hours (George, 1993). *See* PHYTOCHROME.

**day neutral plant:** A plant that flowers regardless of daylength of light received (Gahan, 1984; Hartmann et al., 2001).

**deceleration phase:** Slowing of growth rate following the linear phase of growth of suspension cultures (Scragg, 1991a). *See* GROWTH STAGES.

**deciduous:** Describes the loss by shedding of leaves, petals, and fruits either seasonally or with senescence; plants that are not evergreen (Graham et al., 2003; Mauseth, 2003).

**decontaminate:** *See* SURFACE STERILIZATION.

**dedifferentiation:** Resumption of meristematic activity by, usually parenchyma, cells that have specialized or partially differentiated and revert through modification of subcellular structures (George, 1993). *See* ADVENTITIOUS REGENERATION; DIFFERENTIATION.

doi:10.1300/5648_04

**deficiency:** Loss of one or a series of genes; inadequate availability of nutritional or environmental agent.

**defined medium:** Medium of specified composition (George, 1993, 1996). *See* PLANT TISSUE CULTURE MEDIA.

**defoliants:** Compounds that cause leaf-drop in plants (Basra, 2000). *See* PLANT GROWTH REGULATOR.

**dehumidifier:** Equipment that removes humidity from the atmosphere.

**dehydration:** Removal of water from plant material; removal of water from material being prepared as permanent mounts for microscopy.

**dehydrogenase:** Oxidoreductase enzyme involved in the transfer of electrons from a reduced substrate to an electron acceptor, e.g., nicotinamide adenine dinucleotide (NAD), nicotinamide adenine dinucleotide phosphate (NADP), flavine adenine dinucleotide (FAD) (Gahan, 1984; Metzler, 2003).

**deionized water:** Water from which soluble minerals and some organic salts have been removed after passage through an ion exchange column. *See* DISTILLED WATER; WATER QUALITY.

**deletion:** Act of the loss of either a piece of chromosome or nucleotide(s) from a nucleic acid (Dyer, 1979; Russell, 2002).

**deliquescence:** Gradual autolysis of tissue; absorbance of water by some chemicals.

**demineralization:** Removal of salts and ions from water using an ion exchange column, electrodialysis, or distillation; removal of minerals from tissues during histological preparation, often through treatment with acid. *See* WATER QUALITY.

**de novo:** Anew.

**deoxyribonucleic acid:** DNA; two polynucleotide chains arranged in a double helix in which the antiparallel chains are arranged with their ends in the 5′ to 3′ and 3′ to 5′ directions. The chains are linked by hydrogen bonding between the adjacent base pairs, cytosine–guanine and thymine–adenine (Metzler, 2001).

**deoxyribonucleotide:** Nucleotide found in DNA and comprised of 2-deoxy-D-ribose sugar, phosphate, and one of the bases thymine, cytosine, guanine, or adenine (Metzler, 2001). *See* DEOXYRIBONU-CLEIC ACID.

**derepression:** The alleviation of repression (Russell, 2002).

**dermal:** Relating to the outer layers of tissues—epidermal, hypo-dermal, peridermal (Esau, 1977).

**dermatogen:** That part of the meristem from which the epidermis is formed according to the histogen theory. *See* APICAL MERISTEM; HISTOGEN THEORY.

**desiccate:** To dry; the removal of water, perhaps with an agent such as sodium hydroxide, a desiccant.

**desiccation:** Drying out of tissue. *See* MICROPLANT ESTABLISHMENT.

**desiccation tolerance:** The plant's ability to function while dehydrated (Lerner, 1999; Taiz and Zeigler, 2002). *See* COMPATIBLE SOLUTES.

**detached meristem:** A meristem that is distant from the apical meristem, e.g., an axillary meristem. *See* APICAL MERISTEM.

**detergent:** A cleaning substance made from, e.g., alkyl sulfonate that can act as either an emulsifier or a wetting agent.

**determinate growth:** Growth that is defined and often has a set duration, resulting in the production of a specific structure, e.g., a flower or leaf (Hartmann et al., 2001; Mauseth, 2003). *See* INDETER-MINATE GROWTH.

**determination:** The commitment of a cell or group of cells to develop along a specified pathway (George, 1993; Scharz et al., 2005). *See* INDUCTION MEDIUM.

**Dettol:** A commercial sterilizing solution of chloroxylenol (George, 1993). *See* SURFACE STERILANTS.

**development:** The regulated changes occurring during the evolving of a plant, or part of a plant, to the mature plant, initially from the fertilized egg or totipotent cell or somatic embryo (Howell, 1998).

**deviation:** Changes occurring to the normal development of plant form or function.

**Dewer flask:** A double-walled flask with a vacuum between the two walls in order to reduce heat loss from, or heat transfer to, the liquid contained in the flask. *See* CRYOPRESERVATION.

**dextrose:** Syn. glucose. *See* GLUCOSE.

**diageotropism:** Growth of rhizomes at a right angle to the direction of gravity (Roberts et al., 2002).

**dicamba:** 3,6-Dichloro-2-methoxybenzoic acid; a synthetic auxin with herbicidal properties. It promotes callus growth in vitro. *See* AUXINS.

**3,6-dichloro-2-methoxybenzoic acid:** *See* DICAMBA.

**2,4-dichlorophenoxyacetic acid:** 2,4-D; a synthetic auxin with herbicidal properties; selective weed killer (Basra, 2000); promoter of callus growth in vitro and inducer of xylogenesis (George, 1993). *See* AUXINS; PLANT GROWTH REGULATOR.

**dichlorophenyldimethylurea:** Syn. diuron; a photosynthetic inhibitor, herbicide (Basra, 2000).

**dicot:** Syn. Dicotyledonae; broadleaf flowering plants; subclass of angiosperms with embryos having two cotyledons; woody or herbaceous plants usually possessing a cambium, a primary root developing into a taproot, broad leaves with branched veins, and a ring of stem vascular bundles (Esau, 1977; Heywood, 1993; Maseuth, 2003).

**differentiation:** Development of specialized functions of cells and tissues involving cellular and morphological changes; some specializations are reversible, e.g., in parenchymal cells, but others are not and are considered to be fully differentiated, e.g., phloem and xylem (Esau, 1977; Trigiano and Gray, 2005). *See* DEDIFFERENTIATION.

**differentiation medium:** A medium on which competent cells become determined and undergo organogenesis (Schwarz et al., 2005). *See* EXPRESSION MEDIUM; INDUCTION MEDIUM.

**diffusion:** Movement of ions and molecules from areas of high concentration (high free energy) to areas of low concentration (low free energy), resulting in an equilibrium between the two areas in a closed system.

**diffusion pressure deficit:** Suction pressure. The deficit is the net force causing water to enter a cell, its magnitude being determined by the difference between the osmotic pressure and the turgor pressure (Ridge, 2002). *See* OSMOTIC PRESSURE; TURGOR PRESSURE.

**digestion:** The enzymatic breakdown of complex, often insoluble, material to simple, soluble structures (Metzler, 2001).

**dihaploid:** The product of a cross between double monoploids (Allard, 1999).

**dihybrid:** An organism that is heterozygous for two particular genes (Allard, 1999).

**dihydrozeatin:** A natural cytokinin (George, 1993). *See* CYTOKININS; PLANT GROWTH REGULATOR.

**dimethyl sulfoxide:** DMSO; a cryoprotectant used in the preparation of frozen ultrathin sections; occasionally used in the preparation of culture media; a solvent. *See* CRYOPROTECTANT.

**dimorphism:** Two distinct forms of structures—whole plants, leaves, appendages, tissues, or organelles (Esau, 1977).

**dioecious:** Describes plants in which the male and female reproductive organs are on separate plants (Heywood, 1993; Mauseth, 2003).

**1,3-diphenylurea:** A nonpurine synthetic plant growth regulator with cytokinin activity (George, 1993). *See* CYTOKININS; PLANT GROWTH REGULATOR.

**diplobiontic:** Describes a life cycle in which two separate types of vegetative plants are formed, one haploid and the other diploid (Graham et al., 2003). *See* DIPLOID.

**diploid:** Having two complete (haploid) sets of chromosomes per nucleus and in all somatic cells of an individual; one set is derived maternally and the other paternally. *See* HAPLOID.

**diplontic:** Describes a life cycle in which the diploid phase dominates the haploid phase (Graham et al., 2003; Mauseth, 2003).

**diplontic drift:** The competition that exists between cells of the apical initial zone and underlies the formation of chimeras and solid mutants (Balkema, 1972). *See* CHIMERA.

**diplontic selection:** *See* DIPLONTIC DRIFT.

**diplospory:** Form of apomixis in which the embryo forms directly from the megaspore mother cell. *See* APOMIXIS.

**direct embryogenesis:** Development of embryoids directly on plant tissues and zygotic and somatic embryos without an intervening callus formation (Bajaj, 1995). *See* INDIRECT EMBRYOGENESIS.

**direct organogenesis:** The formation of adventitious shoots from the explant without intervening callus proliferation (George, 1993; Schwarz et al., 2005). *See* SOMACLONAL VARIATION.

**disaccharide:** Two monosaccharides linked by a glycosidic bond, e.g., sucrose (Metzler, 2001).

**disease:** A disorder in plants and tissue cultures induced by either environmental events or pathogens leading to the breakdown of normal physiological and morphological conditions (Lerner, 1999; Taiz and Zeigler, 2002; Strange, 2003).

**disease elimination:** The freeing of plants from pathogens (Hadidi et al., 1998; Cassells, 2000a). *See* ANTIBIOTICS; ANTIVIRAL COMPOUNDS; CHEMOTHERAPY; MERISTEM CULTURE; THERMOTHERAPY.

**disease free:** Describes a plant showing no signs of disease and being pathogen free after repeated tests. *See* PLANT HEALTH CERTIFICATION.

**disease indexing:** The screening of plants for the presence of pathogens (Cassells, 2000a). *See* PATHOGEN INDEXING.

**disease resistance:** Describes a plant that is not colonized by a specific pathogen. Resistance may be based on major genes or polygenes (Strange, 2003). *See* BIOTIC STRESS.

**disinfectants:** *See* SURFACE STERILIZATION.

**disinfection:** The use of either chemical or physical agents to destroy disease-causing viruses, bacteria, or fungi. *See* SURFACE STERILANTS.

**dispense:** To proportion out tissue culture medium, e.g., into containers.

**dissecting microscope:** Low-power microscope (~50 to 100×) with zoom lens facility and a broad field of vision to enable the examination of parts of whole organisms and their dissection (Murphy, 2001).

**dissection:** Cutting out; opening up plant or animal material to display the internal structure and organization.

**dissociation:** Separation of units, e.g., protoplasts from a tissue mass after enzyme hydrolysis of cell walls, ions from a molecule in solution.

**dissolve:** Take up in solution. *See* DIMETHYL SULFOXIDE.

**distal:** Describes the point furthest away from the point of attachment to the plant body.

**distilled water:** Water purified by evaporation followed by condensation. *See* WATER QUALITY.

**dithiothreitol:** A reducing agent that inhibits the synthesis of zeaxanthin. *See* CYTOKININ INHIBITORS.

**diurnal rhythm:** An endogenous daytime response rhythm controlled by a circadian clock (Ridge, 2002). *See* CIRCADIAN RHYTHM.

**DMSO:** *See* DIMETHYL SULFOXIDE.

**DNA:** *See* DEOXYRIBONUCLEIC ACID.

**DNA chips:** *See* DNA MICROARRAYS.

**DNA cloning:** Production of large numbers of identical copies of DNA, usually performed in bacterial cells (Russell, 2002).

**DNA denaturation:** Syn. DNA melting (Russell, 2002). *See* POLYMERASE CHAIN REACTION.

**DNA fingerprinting:** The use of restriction fragments to identify an individual (Karp et al., 1998; Russell, 2002). *See* AMPLIFIED FRAGMENT LENGTH POLYMORPHISM TECHNIQUE; MICROSATELLITES; POLYMERASE CHAIN REACTION; RESTRICTION FRAGMENT LENGTH POLYMORPHISM.

**DNA ligase:** An enzyme that can join segments of DNA together (Karp et al., 1998; Alberts et al., 2002; Russell, 2002). *See* GENETIC ENGINEERING.

**DNA methylation:** A postsynthetic modification of DNA. Of the four base pairs, only cytosine–guanine (CG) is methylated. As DNA replication is semiconservative, one strand of the new DNA is premethylated and the other strand is methylated by a methylase. As not all CG base pairs in DNA are methylated, the methylase enzyme must recognize the appropriate base pair to ensure that both strands of the newly synthesized DNA are identical. Methylation protects the host DNA from endonucleases designed to destroy foreign DNA (mainly bacterial DNA) and acts by blocking endonuclease restriction sites. Methylation also regulates gene expression, with experimental evidence showing that heavily methylated genes are not expressed (Howell, 1998; Russell, 2002). Methylation may be associated with some adaptations to in vitro conditions (Joyce and Cassells, 2002). *See* EPIGENETIC VARIATION.

**DNA microarrays:** An ordered array of DNA molecules of known sequence immobilized on a solid substrate, e.g., a silicon chip. Labeled free DNA molecules added to the fixed probes bind to complementary sequences, allowing identification and quantification of the target DNA (Baldi and Hatfield, 2002).

**DNA probe:** A nucleic acid sequence complementary to a region being sought in a DNA restriction digest (Karp et al., 1998; Russell, 2002). Detection can be based on radioactivity (use of a radiolabeled probe) or by amplification by polymerase chain reaction. *See* BIOTIN LABELING; POLYMERASE CHAIN REACTION; RADIOLABELING.

**DNase:** An enzyme that hydrolyzes DNA (Russell, 2002).

**DNA uptake into protoplasts:** Method used for virus inoculation of protoplasts; polyethylene glycol and calcium can be applied to transform protoplasts with transgenes (Cassells, 1989). DNA can also be introduced into protoplasts using electroporation, liposomes, and biolistic methods. *See* GENE TRANSFER METHODS.

**Domestos:** A commercial sterilizing solution containing hypochlorite (George, 1993). *See* SURFACE STERILANTS.

**dominant trait:** Masking of an allele by the other allele at the same locus; the heterozygous phenotype resembles the plant homozygous for two dominant alleles (Allard, 1999). *See* HETEROZYGOUS; HOMOZYGOUS; RECESSIVE TRAIT.

**donor plant:** Plant source used for propagation (Cassells and Doyle, 2004). *See* MICROPROPAGATION.

**dormancy:** Inactive phase occurring in seeds, spores, buds, and other plant organs (Taiz and Zeigler, 2002).

**double antibody sandwich:** DAS; *see* ENZYME-LINKED IMMUNO-SORBENT ASSAY.

**doubling time:** The time taken for a prokaryote or eukaryote cell to divide (Scragg, 1991a; Prescott et al., 2001).

**drought:** Lack of available water. *See* ABIOTIC STRESS.

**drought resistance:** A plant's ability to restrict damage due to water stress (Basra and Basra, 1997). *See* COMPATIBLE SOLUTES.

**dry ice:** Solid carbon dioxide at a temperature of $-78°C$.

**dry weight:** Weight of plant material after drying either at 60°C in an oven or freeze-drying, so avoiding the variability in fresh weight measurement as a result of fluctuations in tissue water content. *See* FREEZE-DRYING.

**dual culture:** Culture of plant tissue together with another organism, usually an obligate pathogen (Meulemans et al., 1987).

**ectomycorrhizal fungi:** Symbiotic root-sheathing fungi that assist the plant with phosphate acquisition and water uptake. The fungi also assist in forest litter decomposition and recycling. They are used to inoculate forest tree species in vivo (Mukerji and Upadhyay, 1999; Duffy and Cassells, 2003). *See* MYCORRHIZAL FUNGI.

**EDTA:** *See* ETHYLENEDIAMINETETRAACETIC ACID.

**EGTA:** *See* ETHYLENE GLYCOL-BIS (B-AMINO-ETHYL ETHER)-N,N,N',N'-TETRAACETIC ACID.

**electrofusion:** The fusion of two or more protoplasts in an electric field. The protoplasts to be fused have to be in contact before the pulsed electric field is applied. Protoplasts have a net negative surface charge and repel one another. A number of chemical techniques overcome charge repulsion, e.g., the use of polyethylene glycol (PEG), high calcium, or high pH (Veilleux et al., 2005). Dielectrophoresis (where the protoplasts are placed between the electrodes in an alternating current) induces an ion charge separation inside the protoplast, forming a dipole. This has two consequences: the protoplast moves toward the point of highest field strength, and when the cells are in close proximity they form long chains called pearl chains (Bates, 1989). Application of a DC pulse then causes localized membrane destabilization and fusion at the junctions, resulting in the production of single and multiple protoplasts fusions to form homo-, hetero-, and mixed fusions. The frequency of heterofusions, i.e., heterokaryon formations, is of the order of a few percent, necessitating an efficient selection system (Warren, 1991). *See* COMPLEMENTATION SELECTION; FLUORESCENCE-ACTIVATED CELL SORTER.

**electromagnetic spectrum:** The band of electromagnetic radiation, which can be separated into components distinguished by their relative wavelengths. The portion of the spectrum that the human eye can detect is called visible light, between the longer infrared waves and

doi:10.1300/5648_05

the shorter ultraviolet waves. The various types of energy comprising the spectrum are (from longest to shortest) radio, infrared, visible, ultraviolet, X-rays, gamma rays, and cosmic rays (Taiz and Zeigler, 2002).

**electron microscope:** EM; a microscope that functions exactly as a light microscope does, except that it uses a beam of electrons focused by a series of magnets, instead of light, to form the image. EMs give a magnification of 100,000×, compared with a limit of 1,300× for the light microscope. More important, the transmission electron microscope (TEM) has a resolution (ability to distinguish between two point sources) of 0.2 nm, as opposed to the 0.24 μm of the light microscope. The TEM was the first type of electron microscope to be developed in the 1930s; the scanning electron microscope (SEM), which is important for examining surfaces, was commercialized in the mid-1960s (Bozzola and Russell, 1998; Murphy, 2001).

**electrophoresis:** *See* GEL ELECTROPHORESIS.

**electroporation:** A method for the introduction of DNA into protoplasts, cells, and tissues. The cells or tissues are placed in a buffer with the target DNA or plasmid containing the transgene and subjected to high-voltage electrical pulses (Slater et al., 2003). *See* GENE TRANSFER METHODS.

**elicitors:** Compounds that function as a signal to the plant to mobilize its defense mechanisms against pests and pathogens (Denny, 2002; Strange, 2003). They include microbial metabolites and cell wall fragments. The defenses involve the synthesis of compatible solutes, volatile chemical attractants of predators and parasitoids, and various categories of stress proteins. Elicitors have scope to be used as biopesticides, offering a potential advantage over traditional pesticides because they are naturally occurring, are active at low doses, and have no direct toxicity to the plant's natural enemies or other nontarget organisms. *See* BIOTIC STRESS.

**ELISA:** *See* ENZYME-LINKED IMMUNOSORBENT ASSAY.

**elite plants:** Plants selected for their characteristics or high health status (van der Linde, 2000). *See* MICROPLANT QUALITY; PLANT HEALTH CERTIFICATION.

**embryo:** The bipolar structure that develops in the seed after fertilization, consisting of a bud (the plumule), a root (the radicle), and one or more seed leaves (cotyledons) (Esau, 1977; Graham et al., 2003). *See* SOMATIC EMBRYOGENESIS.

**embryogenesis:** The development of the embryo, normally from a fertilized egg cell. In vitro embryogenesis is discussed in Bhojwani and Soh (2001) and Thorpe (2002). *See* SOMATIC EMBRYOGENESIS; ZYGOTIC EMBRYO.

**embryogenic callus:** Callus giving rise to adventitious embryos (Bajaj, 1995). *See* SOMATIC EMBRYOGENESIS.

**embryoids:** *See* SOMATIC EMBRYO.

**embryonal suspensor:** The line of cells that arises from the proembryo by mitosis. It connects the embryo to the parental tissue and conducts nutrients to the embryo (Esau, 1977).

**embryo rescue:** An in vitro technique for the recovery of embryos where the endosperm fails to develop normally as a result of a form of incompatibility in wide crosses (Taji et al., 2001; Cassells, 2002). *See* POSTZYGOTIC INCOMPATIBILITY.

**EMS:** *See* ETHYLMETHYL SULFONATE.

**endomitosis:** Syn. endopolyploidy; DNA and chromatid replication with chromatid separation without a following anaphase, resulting in a doubling of the chromosome number. This may be repeated to yield a high level of polyploidy. It can be experimentally induced by high or low temperature treatment or through radiation (Darlington and La Cour, 1976). *See* POLYPLOIDY.

**encapsulation:** Inclusion in a coating, e.g., the coating of artificial seeds to provide an artificial seed coat (Redenbaugh, 1993; Bajaj, 1995). *See* ARTIFICIAL SEED.

**endocytosis:** Absorption of materials into a cell where the plasmalemma invaginates and a vesicle is pinched off, carrying the material into the cytoplasm, where it is directed either to the endosomal/lysosomal system or to the plasmalemma, where the contents are exocytosed.

**endogenous:** Originating from inside the system. *See* EXOGENOUS.

**endogenous growth regulators:** Plant growth regulators in the explant tissue as opposed to those incorporated into the medium.

**endophytes:** *See* ENDOPHYTIC ORGANISMS.

**endophytic organisms:** Syn. endophytes; microorganisms existing within the tissues of a plant. Endophytes, in contrast to plant pathogens, are not generally associated with disease but can carry over in explants to contaminate cultures. They are generally plant-surface-associated or environmental microorganisms (Cassells and Tahmatsidou, 1997; Nowak et al., 1997; Cassells and Doyle, 2004, 2005). *See* EPIPHYTE; HEMIENDOPHYTE.

**endopolyploidy:** A doubling of the chromosome number resulting from endomitosis (mitosis without cell division; D'Amato, 1997). Endopolyploidy is common in some tissues, e.g., xylem, but may also occur in root cortical cells, sometimes in response to mycorrhizal colonization. Such cells may give rise to polyploid adventitious shoots (Buiatti, 1997). *See* SOMACLONAL VARIATION.

**endosperm:** The storage tissue in the seed of most angiosperms formed from one of the two cells from the first division of the zygote (Graham et al., 2003). *See* ARTIFICIAL SEED.

**endosymbiont hypothesis:** Syn. endosymbiont theory; hypothesizes that plastids and mitochondria arose from symbiotic prokaryote ancestors living in eukaryotic cells (Margulis, 1981; Russell, 2002; Mauseth, 2003).

**environmental stress:** *See* ABIOTIC STRESS.

**enzymatic cell separation:** The separation of plant cells by pectinases, the first stage in protoplast isolation (Warren, 1991). *See* PROTOPLAST ISOLATION.

**enzyme-linked immunosorbent assay:** ELISA; an assay combining the specificity of antibodies with the sensitivity of an enzyme assay, using antibodies or antigens coupled to an easily assayed enzyme. ELISA can be used to detect and quantify the presence of antigens that are recognized by an antibody or it can be used to test for antibodies that recognize an antigen. Double antibody sandwich (DAS) ELISA is a five-step procedure: the microtiter plate wells are coated with antigen; all unbound sites are blocked to prevent false positive results; antibody is then added to the wells; if the primary antibody was raised in mouse, antimouse IgG conjugated to an enzyme is added as the reporter system; substrate to the enzyme is added to produce a colored product, which is quantified and compared with negative and positive controls. Several different types of ELISA are in use (Dijkstra and de Jager, 1998; Martin, 1998). *See* PATHOGEN INDEXING.

**epicormic shoots:** Stump sprouts that retain a juvenile physiological status (George, 1993; Hartmann et al., 2001). *See* REJUVENATION.

**epicotyl:** The part of the seedling stem above the cotyledons (Esau, 1997).

**epicuticular wax:** The surface covering of the aboveground parts of plants, consisting of a cutin matrix and soluble cuticular waxes (Brett and Waldron, 1989; Carpita et al., 2001; Strange, 2003). *See* CUTICLE; INTRACUTICULAR WAX.

**epidermis:** The outermost layer of plant cells; usually one layer (Esau, 1977). *See* APICAL MERISTEM.

**epifluorescence:** An optical accessory for a fluorescence microscope whereby the excitation light passes from the light source through the eyepiece to the specimen. The objective lens is used to focus an excitation wavelength of light onto a specimen containing a fluorochrome, and the fluorescent light from the specimen is collected. Epifluorescence is more efficient than transmitted fluorescence, in which the excitation light passes through a condenser to the specimen (Hawes and Satiat-Jeunemaitre, 2001; Murphy, 2001).

**epifluorescence microscopy:** *See* FLUORESCENCE MICROSCOPY.

**epigenetics:** The study of heritable changes in gene function that occur without a change in the DNA sequence (Anonymous, 1998; Russo et al., 1999; van de Vijver and Waele, 2002).

**epigenetic variation:** Variation resulting from altered gene expression as opposed to mutation. Epigenetic variation is associated with adaptation to in vitro conditions and may result in, e.g., a directed change in leaf shape, altered disease susceptibility, prolonged juvenility, or delayed flowering (Cassells and Morrish, 1987; Cassells et al., 1999; Rival et al., 2000). Epigenetic variation is a component of somaclonal variation (Mohan Jain et al., 1998). *See* SOMACLONAL VARIATION.

**epinasty:** The downward bending of leaves resulting from different growth rates on the upper and lower sides in response to ethylene production during flooding (Taiz and Zeigler, 2002).

**epiphyte:** An organism growing on another organism. The term can refer to microorganisms living on the surfaces of plant tissues (Campbell, 1989). *See* ENDOPHYTIC ORGANISMS; HEMIENDOPHYTE.

**epistasis:** The control of expression of one gene by another. Epistasis can be an issue in genetic engineering (Russell, 2002; Slater, 2003). *See* GENETIC ENGINEERING.

**EPPO:** *See* EUROPEAN AND MEDITERRANEAN PLANT PROTECTION ORGANIZATION.

**erythromycin:** A macrolide antibiotic produced by a strain of *Streptomyces erythraeus* (Walsh, 2003). *See* ANTIBIOTICS.

**escapes:** Plants that have escaped infection by pathogens. They are sought as donor plants for tissue culture (Cassells and Doyle, 2004).

**essential elements:** Minerals essential for normal plant growth and development (Marschner, 1994). *See* MINERAL NUTRITION.

**establishment:** The achievement of propagule growth in the greenhouse or field. *See* MICROPLANT ESTABLISHMENT.

**ethane:** A gas. Ethane is an indicator of lipid peroxidation, indicating severe cell damage (Inze and van Montague, 2001); it has been reported as a predictive parameter of protoplast survival potential (Cassells and Tamma, 1985). *See* PROTOPLAST ISOLATION.

**ethanol:** Ethyl alcohol; used in flame sterilization of instruments and as a fixative and dehydrating agent in histology and cytology. *See* INSTRUMENT STERILIZATION.

**Ethephon:** Syn. Ethrel; an ethylene-releasing agent used commercially as a defoliant (Basra, 2000). *See* ETHYLENE.

**Ethrel:** *See* ETHEPHON.

**ethylene:** Syn. ethene; a plant growth regulator. Ethylene is produced from methionine in all tissues in higher plants. Ethylene research has focused on the synthesis-promoting effects of auxin, wounding, drought resistance, and fruit ripening. ACC synthase is the rate-limiting step for ethylene production and has been genetically manipulated to delay fruit ripening in the "flavor saver" tomatoes (Slater et al., 2003). Ethylene stimulates the following plant processes: the release of dormancy; shoot and root growth and differentiation (triple response); leaf and fruit abscission; flower induction in bromeliads; flower opening; flower and leaf senescence; ripening plus the induction of femaleness in dioecious flowers; possibly, adventitious root formation (Taiz and Zeigler, 2002). A number of ethylene-releasing compounds are used in horticulture, e.g., Ethrel, Ethephon (Basra, 2000). Accumulation of ethylene in vitro is associated with aberrant morphology (Cassells et al., 2003). *See* ETHEPHON; ETHYLENE INHIBITORS.

**ethylenediaminetetraacetic acid:** EDTA; a chelating agent. *See* CHELATING AGENTS.

**ethylene glycol-bis (b-amino-ethyl ether)-N,N,N′,N′-tetraacetic acid:** EGTA; a chelating agent. *See* CHELATING AGENTS.

**ethylene inhibitors:** Aminoethoxyvinylglycine (AVG), aminooxy-acetic acid (AOA), and cobalt are inhibitors of ethylene synthesis. Silver ions (used as silver nitrate or silver thiosulfate) are inhibitors of ethylene action and are frequently incorporated in plant tissue culture media. *Trans*-cyclooctene and 1-methylcyclopropane are volatile inhibitors of ethylene binding to its receptor. Commercial ethylene binding agents, e.g., Ethysorb, are available and are sometimes included in culture vessels (George, 1993; Taiz and Zeigler, 2002). *See* ETHYLENE; ETHYLENE RECEPTORS; HYPERHYDRICITY; TISSUE CULTURE VESSELS.

**ethylene receptors:** The first ethylene receptor, *etr1,* was identified from a screening of ethylene-insensitive *Arabidopsis* mutants. Subsequently, other receptors have been identified. The receptors consist of a histidine kinase binding site and response regulator that may function as a transcription factor (Taiz and Zeigler, 2002). *See* ETHYLENE INHIBITORS.

**ethylene triple response:** A common response to ethylene involving a reduction of the rate of elongation, increased lateral expansion, and swelling in the region below the hook in etiolated seedlings of most dicots and of coleoptiles and mesocotyls of oat and wheat seedlings (Taiz and Zeigler, 2002). *See* BIOASSAY.

**ethylmethyl sulfonate:** EMS; a chemical mutagen (van Harten, 1998). *See* CHEMICAL MUTAGENS.

**Ethysorb:** A commercial ethylene absorber based on aluminum oxide and potassium permanganate. Ethysorb has been used in culture vessels to bind ethylene (George, 1993). *See* ETHYLENE TISSUE CULTURE VESSELS.

**etiolation:** A form of growth seen in plants receiving insufficient light. It is characterized by long, weak stems, small leaves, and chlorosis (Ridge, 2002).

**euchromatin:** The part of the chromosome at interphase that is relatively unfolded to permit gene transcription (Alberts et al., 2002). *See* HETEROCHROMATIN.

**eukaryotes:** Organisms with cells containing a membrane surrounding the nucleus, in which the DNA is complexed with histones into chromosomes. Eukaryotic cells have a cytoskeleton of filaments and tubules. Many processes are compartmentalized in organelles. Of the latter, the mitochondrion and chloroplast have bacterial-like genomes and ribosomes. It is widely accepted (the endosymbiont hypothesis) that eukaryote cells arose from fusions between free-living anaerobic, aerobic, and photosynthetic bacteria (Margulis, 1981; Mauseth, 2003). *See* ENDOSYMBIONT HYPOTHESIS; PROKARYOTE.

**euploid:** Having an exact multiple of the haploid number of chromosomes, e.g., diploid, triploid, tetraploid (Dyer, 1979). *See* ANEUPLOID.

**European and Mediterranean Plant Protection Organization:** EPPO; an intergovernmental organization responsible for European cooperation in plant protection in the European and Mediterranean region. Under the International Plant Protection Convention (IPPC), EPPO is the regional plant protection organization (RPPO) for Europe (Krczal, 1998). *See* PLANT HEALTH CERTIFICATION.

**excision:** The cutting out, e.g., of tissues. *See* EXPLANT.

**exogenous:** Originating from outside the system. *See* ENDOGENOUS.

**explant:** A piece of excised tissue used to establish a plant tissue culture. *See* MICROPROPAGATION.

**exponential growth:** *See* GROWTH STAGES.

**expression medium:** A medium on which cells dedifferentiate, enter a competent state, become determined, and undergo organogenesis (Schwarz et al., 2005). *See* COMPETENCE; DIFFERENTIATION MEDIUM; INDUCTION MEDIUM.

**extra vitrum:** Lit. "outside glass"; used to refer to the stages after microshoots or microplants have been removed from the culture vessel.

**exudate:** Leakage from an organism, tissue, or cell.

**ex vitro:** *See* EXTRA VITRUM.

**F₁ generation:** The heterozygous offspring resulting from a cross between two parents homozygous for contrasting alleles (Allard, 1999).

**FACS:** *See* FLUORESCENCE-ACTIVATED CELL SORTER.

**factor:** Inherited component responsible for the determination of a characteristic, i.e., a gene.

**factorial experiment:** An experiment divided in such a way that combinations of treatments can be tried individually on at least one sample of, e.g., callus (Motulsky, 1995; Compton, 2005). *See* RANDOMIZED BLOCK DESIGN.

**facultative:** Having the ability to exploit particular circumstances or environments without being dependent upon them. Many soil-inhabiting fungi are facultative pathogens (Strange, 2003) of microplants (Williamson et al., 1997).

**FAD:** *See* FLAVINE ADENINE DINUCLEOTIDE.

**falcate:** Sickle-shaped; used, e.g., of leaves.

**FAO:** *See* FOOD AND AGRICULTURE ORGANIZATION OF THE UNITED NATIONS.

**FAP:** *See* FATTY ACID PROFILING.

**far-red light:** Electromagnetic radiation of a wavelength of above 740 nm. *See* ELECTROMAGNETIC SPECTRUM.

**far-red light responses:** Far-red light is perceived by phytochromes. A red-light-sensitive form maximally absorbs red light at 660 nm and a far-red light form maximally absorbs far-red light at 730 nm. The switching between red and far-red light forms is associated with a

doi:10.1300/5648_06

number of physiological events, including stem elongation, seed germination, and floral induction (Taiz and Zeigler, 2002). *See* ARTIFICIAL LIGHT; GROWTH ROOM LIGHTING; PHYTOCHROME.

**fasciation:** An abnormal flattening or coalescence of stem stalks.

**fascicular cambium:** Vascular cambium originating within the vascular bundle (Esau, 1977). *See* VASCULAR CAMBIUM.

**Fast Green:** Dye used histologically to stain cytoplasm and cellulose structures and cytochemically to react specifically at pH 9.0 with histones after the removal of nucleic acids from sections or cells (Gahan, 1984).

**fastidious bacteria:** Bacteria that have special nutrient requirements not provided by plant tissue culture media (Prescott et al., 2001). *See* LATENT CONTAMINATION; MOLLICUTES.

**fastidious contaminants:** Contaminants of tissue cultures and plants that have specific culture media requirements; contaminants that do not grow on common culture media.

**fatty acid:** Long chain aliphatic acid that may be saturated or unsaturated, present in phospholipids. The degree of unsaturation affects the fluidity of membranes in that a "kink" occurs in the straight fatty acid chain of the phospholipid, resulting in the increased spacing of the molecules and permitting rotation and lateral movement, leading to increased fluidity of the membrane (Metzler, 2001; Alberts et al., 2002).

**fatty acid profiling:** FAP; application of fatty acid analysis to the identification of a microorganism. FAP involves the following steps: culture of the microorganism under standard conditions, extraction of the fatty acids from the cell surface by saponification, methylation of the fatty acids to increase volatility, and analysis using gas chromatography. The identification of microorganisms is based on a comparison of the fatty acid profile of an unknown to those in the literature or by computer library matching (Ackman and Metcalfe, 1976; Stead et al., 1992, 2000). *See* BACTERIAL IDENTIFICATION; DNA FINGERPRINTING.

**FDA:** *See* FLUORESCEIN DIACETATE.

**Fe:** *See* IRON.

**feedback inhibition:** Inhibition by the product of a sequence of reactions, as when the product of a multienzyme pathway inhibits an enzyme earlier in the pathway, e.g., inhibition of phosphofructokinase by citrate (Metzler, 2001).

**feeder cells:** Cells cocultivated with, and providing nutrients and other components for, target cells or protoplasts (Warren, 1991). *See* NURSE CULTURE; PROTOPLAST CULTURE.

**feeder layer:** A layer of feeder cells overlaid with the target cells (Warren, 1991). *See* PROTOPLAST CULTURE.

**Fe-EDTA:** Ferric ethylenediaminetetraacetate; chelated iron. *See* CHELATING AGENTS; IRON.

**fermentation:** Breakdown of glucose under anaerobic conditions, frequently by glycolysis, to yield usable energy (Taiz and Zeigler, 2002). A range of other fermentations are known for bacteria and microorganisms (Parekh and Vinci, 2003).

**fermenter:** Apparatus used for somatic embryogenesis, micropropagation and secondary metabolite production on either a pilot or industrial scale (Scragg, 1991a; Hvoslef-Eide and Preil, 2004). *See* BIOREACTOR.

**ferric ethylenediaminetetraacetate, sodium salt:** Fe Na-EDTA; used in some tissue culture media as a source of iron and sodium (Metzler, 2001).

**fertilization:** Fusion of, especially the nuclei of, haploid male and female gametes to form a diploid zygote during sexual reproduction (Graham et al., 2003; Mauseth, 2003).

**fertilizers:** Chemical nutrients that plants need for growth. The most important are carbon, hydrogen, and oxygen from air and water and the macronutrients nitrogen, phosphorus, and potassium (syn. potash) generally applied to crops, which may also require sulfur, calcium, and magnesium, depending on soil analysis. The micronutrients required are boron, cobalt, copper, iron, manganese, molybdenum, and zinc (Marschner, 1994). *See* MINERAL NUTRITION.

**Feulgen reaction:** Quantitative cytochemical reaction for DNA in which acid hydrolysis removes purines and opens the deoxyribose sugar ring, forming aldehyde groups that react to give a magenta color with Schiff's base (decolorized pararosanilin). The intensity of color is directly proportional to the amount of DNA (Gahan, 1984). *See* FLOW CYTOMETRY.

**fiber:** Rather long, simple-pitted, lignified sclerenchyma cell derived from meristematic cells (Esau, 1977).

**fiber optics:** Fibers of glass, usually about 120 μm in diameter, that are used to carry light or signals in the form of pulses of light.

**ficoll:** Inert, synthetic, highly soluble polymer used as an osmoticum for, e.g., suspending protoplasts (Hunter, 1993).

**field trial:** Assessment of the performance of experimental material in the field.

**filament:** Stalk of the stamen bearing the anther (Esau, 1977); single file of cells of some organisms, e.g., algae (Mauseth, 2003).

**filial generation:** *See* $F_1$ GENERATION.

**filter paper:** A range of porous papers with different sized pores used in filtration and in some forms of tissue culture. *See* TISSUE SUPPORTS.

**filter paper bridges:** Paper tissue supports, as alternatives to gelling agents, used as wicks to convey nutrients to cells and tissues in culture (George, 1993). *See* TISSUE SUPPORTS.

**filters:** Materials that remove particles and microorganisms on the basis of size. *See* ULTRAFILTRATION.

**filter-sterilization:** Passage of solution through a sterile, porous material that removes microorganisms and their spores to render the material sterile. Filter-sterilization is used, e.g., when medium contains heat-labile components (George, 1993; Beyl, 2005). *See* MEDIA PREPARATION.

**filtration:** Passage of either a solution through porous material to remove solids or cultures through meshes of different pore sizes to remove cell aggregates. *See* ULTRAFILTRATION.

**FITC:** *See* FLUORESCEIN ISOTHIOCYANATE.

**fixation:** Attainment of frequency of 100 percent by an allele in a population as a result of complete loss of all other allelic forms of the gene (Allard, 1999); using a solution or vapor of various chemicals to maintain subcelluar structures and tissue forms to further process material for various forms of microscope study of either the histology or cellular or subcellular structures within (Gahan, 1984). *See* FIXATIVE.

**fixative:** The chemical, or combination of chemicals, used in fixation to stabilize the tissues and cellular components in as lifelike a manner as possible, e.g., paraformaldehyde (Gahan, 1984). *See* FIXATION.

**flame-sterilization:** Sterilization of instruments used in tissue culture by heating them in a flame until red and then cooling in a sterile environment. *See* BACTICINERATOR; INSTRUMENT STERILIZATION.

**flank meristem:** Meristem giving rise to the cortical tissues of the plant (See Figure 8). See RIB MERISTEM.

**flat blade impeller:** Propellers used to stir bioreactors (Scragg, 1991a). *See* BIOREACTOR.

**flavanoids/flavonoids:** The most widely distributed group of plant secondary compounds, containing a 2-phenylbenzopyran nucleus and based upon a 15-C atom skeleton of flavone. They are sap-soluble pigments present mainly in combination with sugar residues and include flavones, anthocyanidins, flavanones, and flavonols, which differ from each other through the nature of the central oxygen-containing ring, the anthocyanidins being oxonium salts (Vickery and Vickery, 1981; Hendry, 1993; Metzler, 2003). *See* ANTHOCYANINS.

**flavanones:** *See* FLAVANOIDS.

**flavin:** Yellow plant pigment with an absorption peak at ~370 nm (Hendry, 1993).

**flavine adenine dinucleotide:** FAD; a riboflavin-derived coenzyme whose prosthetic group acts as an electron acceptor for many dehydrogenases, e.g., succinic dehydrogenase (Metzler, 2003).

**flavine mononucleotide:** FMN; a riboflavin-derived coenzyme whose prosthetic group acts as an electron acceptor for many dehydrogenases, e.g., NADH dehydrogenase (Metzler, 2001).

**floccule:** Aggregation of colloidal particles or microorganisms floating in or on a liquid.

**floral meristem:** Forms the flower or reproductive organs (usually, sepals, petals, stamens, carpels) either directly from a vegetative meristem or via an inflorescence meristem (Howell, 1998; Taiz and Zeigler, 2002; Mauseth, 2003). *See* APICAL MERISTEM.

**flow cytometry:** In microscopy a sample is placed on a microscope slide and the objective lens ("detector") is moved over the sample; in contrast, in the flow cytometer the detectors are fixed in position and the sample is moving. At the measuring point the stream of cells intersects a beam of light. In most flow cytometers the light source is a laser or an arc lamp. A detector analyzes the sample, which is typically stained for DNA. Flow cytometers are used to count cells or the percentage of viable cells and to measure DNA content (ploidy analysis, detection of aneuploids; Curry and Cassells, 1998; Givan,

2001), protein and RNA content, and enzyme activities based upon reactions with specific fluorochromes.

**flower:** Sexual reproductive structure of angiosperms containing calyx, corolla, androecium, and gynaecium, though not all of these may be present in all species (Esau, 1977; Graham et al., 2003; Mauseth, 2003).

**flowering plants:** *See* ANGIOSPERMAE.

**fluid drilling:** The sowing of pregerminated seeds in a protective fluid carrier. Emergence of the seedlings has been shown to be much more rapid in the majority of vegetable crops studied, and the number of seedlings established has been more predictable (Hartmann et al., 2001).

**fluorescein diacetate:** FDA; nonfluorescent molecule. In the presence of plasma membranes of living cells or protoplasts, FDA is acted on by an esterase that cleaves the acetate to leave fluorescein, which passes into the cell or protoplast and fluoresces to indicate viability. Dead cells and protoplasts lack this enzyme (Gahan, 1989; Hunter, 1993).

**fluorescein isothiocyanate:** FITC; fluorescent molecule with a variety of uses, including tagging of antibodies and of protoplasts to be fused with unlabeled protoplasts (Gahan, 1989; Hunter, 1993).

**fluorescence-activated cell sorter:** FACS; a flow cytometer adapted to recognize parental and fused cells in droplets based on their fluorescence characteristics. It applies a positive, negative, or no charge, as appropriate, to the droplets, which pass through a magnetic field and are separated on the basis of their charge properties (Warren, 1991; Hunter, 1993). *See* FLOW CYTOMETRY.

**fluorescence microscopy:** A form of quantitative optical microscopy employing light of a selected wavelength to excite specific dye compounds (e.g., acridine orange, DAPI) or primary fluorochromes (e.g., chlorophyll). The energy emitted from such molecules is at a wavelength displaced toward the red end of the spectrum; excitation wavelengths may be from an ultraviolet lamp, white light source, or laser source. Fluorochromes linked to antibodies can be used to stain specific structures, including viruses, in the cell (Gahan, 1984; Hari and Das, 1998; Murphy, 2001).

**fluorescent dyes:** Colorless or yellow-orange molecules that fluoresce in the presence of a particular excitation wavelength (Gahan,1984).

**flux density:** *See* PHOTON FLUX DENSITY.

**FMN:** *See* FLAVINE MONONUCLEOTIDE.

**foam plug:** Autoclavable, reusable sponge stopper.

**foam tissue support:** *See* TISSUE SUPPORTS.

**fog:** Fine particles of liquid suspended in air used to maintain high humidity during plant establishment (Hartmann et al., 2001). *See* MICROPLANT ESTABLISHMENT.

**fogging:** Using a fine mist to irrigate plants in vegetative propagation in horticulture (Hartmann et al., 2001). *See* MICROPLANT ESTABLISHMENT.

**foliage plants:** Ornamental plants grown for their attractive foliage (Hartmann et al., 2001).

**folic acid:** Syn. pterylglutamic acid (Metzler, 2003); a vitamin B complex occasionally added to culture media (George, 1993, 1996).

**Food and Agriculture Organization of the United Nations:** FAO; an agency whose mandate is "to raise levels of nutrition, improve agricultural productivity, better the lives of rural populations and contribute to the growth of the world economy." The FAO establishes guidelines for the international movement of plant material. *See* EUROPEAN AND MEDITERRANEAN PLANT PROTECTION ORGANIZATION; NORTH AMERICAN PLANT PROTECTION ORGANIZATION; PLANT HEALTH CERTIFICATION.

**foot candle:** ~10 lux (an obsolete term).

**forceps:** Long, handheld, metal pincerlike tool.

**formaldehyde:** A preservative and histological fixative. The concentrated form is ~36 percent w/v. *See* FORMALIN.

**formalin:** 100 percent formalin is equivalent to 36 percent w/v formaldehyde. *See* FORMALDEHYDE.

**formazan:** Colored, reduced form of a tetrazole salt. Some formazans form very small crystals, making them good cytochemical agents for studying dehydrogenases, e.g., nitroblue tetrazolium formazan. Others tetrazoles have very soluble formazans, which makes them useful for biochemical assays (Gahan, 1984). *See* TETRAZOLIUM.

**formula:** Prescribed form, method, or recipe.

**formula weight:** Gram molecular weight of a chemical.

**free radicals:** Unstable atoms or molecules that are highly reactive and short lived (Favier et al., 1995). Free radicals are produced naturally during oxidative stress (Inze and van Montague, 2001) and by ionizing radiation; they are missing one or more electrons and are mutagens (Prasad, 1995; van Harten, 1998): as they try to gain the missing electron(s), they can cause damage to proteins and nucleic acids. They have been implicated in somaclonal variation (Cassells and Curry, 2001; Gaspar et al., 2002). *See* OXIDATIVE STRESS; PHYSICAL MUTAGEN; SOMACLONAL VARIATION.

**freeze-drying:** Lyophilizing; allowing frozen material to dry by sublimation at a low temperature.

**freeze-etching:** A modification of freeze-fracturing in which the surface is allowed to lose water prior to making a replica (Bozzola and Russell, 1998; Murphy, 2001). *See* FREEZE-FRACTURING.

**freeze-fracturing:** An electron microscopy preparative method in which the frozen specimen is cleaved along lines of weakness prior to coating with carbon or platinum. The layer is floated off as a replica of the surface structure and examined by electron microscopy (Bozzola and Russell, 1998; Murphy, 2001). *See* FREEZE-ETCHING.

**freeze preservation:** *See* CRYOPRESERVATION.

**fresh weight:** Weight of natural plant material.

**friable:** Readily separable; used, e.g., of cells in a callus (George, 1993).

**fructose:** Monosaccharide ketosugar component of sucrose occurring free in plants (Metzler, 2001). *See* MONOSACCHARIDES; SUCROSE.

**fruit:** Loosely used to describe the fleshy structures associated with gymnosperm seeds; primarily used to describe the structure formed from the pericarp (ovary wall) as the enclosed angiosperm seeds mature (Esau, 1977; Heywood, 1993).

**Fuch Rosenthal chamber:** A chamber for counting cells or protoplasts with a microscope (Hunter, 1993).

**fuchsin:** Dye used in classical chromosome studies (Darlington and La Cour, 1976).

**fume hood:** Enclosed cabinet with air extraction for the handling of hazardous chemicals.

**fumigation:** Exposure to fumes for the purpose of disinfection.

**fungal identification:** Commonly based on morphology (Alexopoulos et al., 1995; Barnett and Hunter, 1998). However, as with bacteria, confirmation of pathogenicity may require inoculation of a susceptible host. A limited number of commercial serological kits are available. Polymerase chain reaction is increasingly being used and offered as a service (Strange, 2003). Filamentous fungi such as *Penicillium* spp. and yeasts are common laboratory contaminants (Leifert and Cassells, 2001; Cassells and Doyle, 2004). *See* FUNGICIDES; GOOD LABORATORY PRACTICE; YEASTS.

**fungicides:** Chemicals that kill fungi (Hewitt, 1998; Strange, 2003) and are occasionally used in tissue culture (George, 1993). *See* ANTIBIOTICS; ANTIMYCOTICS.

**fungus, pl. fungi:** A group of saprophytic, parasitic, or symbiotic eukaryotes that lack chlorophyll and are hence heterotrophic. The cell walls are composed of either chitin or fungal cellulose (Agrios, 1997; Watkinson and Gooday, 2001; Strange, 2003). *See* EUKARYOTES; MYCORRHIZAL FUNGI.

**funiculus:** Stalk attaching the ovule to the ovary wall in angiosperms (Esau, 1977).

**6-furfurylaminopurine:** *See* KINETIN.

**fusaric acid:** A phytotoxin produced mainly by *Fusarium moniliforme* and also by other species of *Fusarium* (Strange, 2003).

**fusicoccin:** An activator of the $H^+$-ATPase thus promoting cellular growth and many transport processes across the plant plasma membrane (Taiz and Zeigler, 2002).

**fusion:** Merging of (1) gametes to form zygotes, (2) cells to form hybrids, (3) protoplasts to from cybrids, and (4) other cell structures to form fusion products, e.g., protoplasts and mitochondria. *See* PROTOPLAST FUSION.

**fusogen:** An agent used to aid somatic hybridization, e.g., polyethylene glycol in protoplast fusion (Warren, 1991; Dodds and Roberts, 1995).

**GA₃:** *See* GIBBERELLIC ACID.

**galactose:** An aldohexose sugar. Not normally in the free state, galactose is a component of the oligosaccharides raffinose and stachyose and of structural polymers, e.g., the galactans with arabinose. Pectic acid is a polymer of oxidized galactose, glucuronic acid (Vickery and Vickery, 1981; Bryant et al., 1999). *See* PECTIN.

**Gamborg's B5 medium:** A medium developed for the culture of soybean root cells now widely used in plant tissue culture (Gamborg et al., 1968).

**gametoclonal variation:** Variation in the progeny of the regenerants from pollen or anther culture (Mohan Jain et al., 1998).

**gametophytic incompatibility:** Incompatibility because of the pollen nucleus (Grant, 1975; de Nettancourt, 1993). *See* POSTZYGOTIC INCOMPATIBILITY.

**gamma radiation:** High-energy radiation used to induce mutation (van Harten, 1998). *See* ELECTROMAGNETIC SPECTRUM; MUTATION BREEDING.

**gaseous permeability of plastics:** Plastics vary in their permeability to plant volatiles such as ethylene and to water, carbon dioxide, and oxygen (Hopfenberg and Stannett, 1974; Neogi, 1996; Massey, 2002). A range of plastic materials, including polycarbonate and polyvinylchloride, are used either to make tissue culture vessels or as lids for culture vessels; the choice of plastic can selectively modify the culture vessel atmosphere (Cassells and Roche, 1994; Cassells and Walsh, 1994; Cassells et al., 2003). *See* TISSUE CULTURE VESSELS.

**gas-liquid chromatography:** Chromatography involving a sample being vaporized and injected onto the head of the chromatographic

doi:10.1300/5648_07

column. The sample is carried through the column by the flow of an inert gaseous mobile phase. The column itself contains a liquid stationary phase, which is adsorbed onto the surface of an inert solid. Separation is based on affinity for and movement with the gas phase. Identification of compounds may involve interfacing the gas chromatograph with a mass spectrometer (GC-MS) (McNair and Miller, 1997; Grob and Barry, 2004). *See* MASS SPECTROMETER.

**gas permeability:** Permeability of the culture vessel to gases. In tissue culture the important gases are oxygen, carbon dioxide, water vapor, and also ethylene. *See* TISSUE CULTURE VESSELS.

**GC-MS:** *See* GAS-LIQUID CHROMATOGRAPHY; MASS SPECTROMETER.

**gelatin:** A gelling agent prepared by the thermal denaturation, with either very dilute acid or alkali, of collagen isolated mainly from animal skin and bones. Gelatin is a heterogeneous mixture of single- and multistranded polypeptides. Type A gelatin (acid pretreatment) is derived from pigskin, whereas Type B gelatin (alkaline treatment) is from cattle hides and bones (Imeson, 1999). *See* TISSUE SUPPORTS.

**gel electrophoresis:** A technique used for the separation of nucleic acids and proteins. Separation of the molecules takes place in a gel in an electric current and depends on the charge and size of the molecules, on the gel pore size, the running buffer, and the temperature. The gel acts as a molecular sieve, where the pore size can be varied by altering the composition of the gel, separating the molecules by size. During electrophoresis, molecules are forced to move through the pores when the electrical current is applied. After staining with, e.g., Coomassie blue for proteins or ethidium bromide for nucleic acids, the separated molecules in each lane can be seen as a series of bands spreading from one end of the gel to the other (Birren and Lai, 1993; Hames and Rickwood, 2001; Westermeier and Barnes, 2001).

**gellan gum:** A bacterial exopolysaccharide from *Sphingomonas elodea*. Gel rigidity depends on the degree of acylation and the ions present. Unacylated gellan forms soft, elastic, transparent, and flexible gels; deacylated gellan forms hard, nonelastic, brittle gels (Imeson, 1999). *See* GELLING AGENT.

**gelling agent:** A jelly, or the solid or semisolid phase of a colloidal solution; a colloidal solution that has set to a jelly such that its viscosity is so great that it has the elasticity of a solid (Imeson, 1999). *See* TISSUE SUPPORTS.

**Gelrite:** A gelling agent composed of glucose, glucuronic acid, and rhamnose residues. Gel strength is determined by choice and concentration of divalent cations ($Mg^{2+}$, $Ca^{2+}$). *See* TISSUE SUPPORTS.

**gene:** A nucleic acid base sequence that functions as or codes for an RNA molecule, a polypeptide, or both (Russell, 2002). *See* GENETIC CODE.

**gene amplification:** Amplification of the number of copies of a gene in the genome (Russell, 2002).

**gene banks:** Typically, collections of seeds from plant varieties and species closely related to crop plants. Gene banks may also contain in vivo, in vitro slow growing, and cryopreserved germplasm of vegetatively propagated crops (Herzberg et al., 1995; Ford-Lloyd et al., 1997; Towill, 2005). *See* GERMPLASM CONSERVATION.

**gene gun:** *See* PARTICLE GUN.

**gene insertion:** *See* GENETIC ENGINEERING.

**gene mutation:** Heritable change of the genetic material, commonly from one allelic form to another (Russell, 2002). *See* CHEMICAL MUTAGENS.

**gene pool:** *See* CROP GENE POOL.

**generative nuclei:** The two male gametes formed by the division of the generative cell in the pollen tube. One of the generative nuclei fuses with the egg nucleus to form the zygote, and the other fuses with the polar nuclei or definitive nucleus to form the primary endosperm nucleus (Graham et al., 2003). *See* FERTILIZATION.

**gene segregation:** *See* SEGREGATION.

**gene silencing:** Instability of transgene expression. At least three mechanisms are found: (1) transcriptional gene silencing involves methylation of homologous sequences in the promoter region, resulting in repression of transcription; (2) posttranscriptional gene silencing is caused by high-level expression of the transgene as a result of a strong promoter activating a mechanism that results in the breakdown of the corresponding transgene mRNA; (3) incorporation of the gene into heterochromatin (Matzke and Matzke, 2000; Hannon, 2003). *See* GENETIC ENGINEERING; RNA INTERFERENCE.

**genetically modified organisms:** GMOs; genetically engineered plants (Slater, 2003). *See* GENETIC ENGINEERING.

**genetic code:** The deoxyribonucleotide triplets that code for the respective protein amino acids and for stop and start codons in DNA. The code is redundant; i.e., each amino acid is specified by more than one triplet (Alberts et al., 2002; Russell, 2002).

**genetic conservation:** *See* GERMPLASM CONSERVATION.

**genetic engineering:** Introduction of a characterized gene or genes other than by conventional breeding. This can be achieved indirectly by *Agrobacterium* transformation or by direct gene transfer using, e.g., particle bombardment ("biolistics") (Slater et al., 2003; Figure 12). *See AGROBACTERIUM*-MEDIATED GENE TRANSFER; GENE TRANSFER METHODS; PARTICLE BOMBARDMENT.

**genetic fingerprinting:** The pattern of DNA fragments of highly variable repeat sequences within the genome, obtained from hydrolysis with restriction enzymes. The number and sizes of the repeat sequences generate unique banding patterns (Karp et al., 1998; Russell, 2002). *See* PATHOGEN INDEXING.

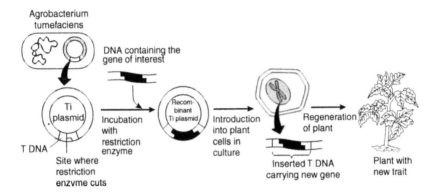

Agrobacterium tumefaciens

DNA containing the gene of interest

Ti plasmid

T DNA

Site where restriction enzyme cuts

Incubation with restriction enzyme

Recombinant Ti plasmid

Introduction into plant cells in culture

Regeneration of plant

Inserted T DNA carrying new gene

Plant with new trait

FIGURE 12. The steps in *Agrobacterium*-mediated plant transformation, involving the isolation of the Ti plasmid from the bacterium, the introduction of the transgene and its promoter, a selectable marker and a reported gene into a modified plasmid. The modified plasmid is multiplied and reintroduced into *Agrobacterium* which is cocultivated with the target plant tissues. The tissues are grown on antibiotic-containing medium, which inhibits nontransformed cells. The tissues are induced to undergo Organogenesis. Expression of the reported gene confirms transformation (Figure 20.19, p. 398 from *Biology, 6th ed.* by Neil A. Campbell and Jane B. Reece. Copyright © 2002 by Pearson Education, Inc. Reprinted by permission.). *See* AGROBACTERIUM; GENETIC ENGINEERING; REPORTER GENE; SELECTABLE MARKER; TRANSGENE.

**genetic manipulation:** Usually refers to methods of plant improvement other than conventional plant breeding (Cassells and Jones, 1995; Cassells and Doyle, 2003). *See* MUTATION BREEDING; PROTOPLAST FUSION.

**genetic stability:** The absence or low frequency of mutations. Although genes mutate naturally, mutation rates are low: of the order of 1 in 500,000 in eukaryotes (Russell, 2002). However, in some genotypes in vitro mutation rates of 10 percent or more have been reported (Mohan Jain et al., 1998). *See* EPIGENETIC VARIATION; SOMACLONAL VARIATION.

**genetic transformation:** *See* GENETIC ENGINEERING.

**genetic variation:** Heritable variation resulting in the presence of different alleles in a population (Russell, 2002).

**gene transfer:** *See* GENETIC ENGINEERING.

**gene transfer methods:** In addition to indirect gene transfer via *Agrobacterium tumefaciens,* four methods for direct gene transfer are established: particle bombardment, electroporation, DNA uptake into protoplasts, and silicon-carbide-fiber-mediated transformation (Slater et al., 2003; Li and Gray, 2005). *See AGROBACTERIUM*-MEDIATED GENE TRANSFER; *AGROBACTERIUM RHIZOGENES; AGROBACTERIUM TUMEFACIENS;* DNA UPTAKE INTO PROTOPLASTS; ELECTROPORATION; SILICON-CARBIDE-FIBER-MEDIATED TRANSFORMATION; PARTICLE BOMBARDMENT.

**gene vector:** A plasmid or nucleic acid carrier of genes for insertion into cells for purposes of genetic engineering (Russell, 2002; Slater, 2003). *See AGROBACTERIUM*-MEDIATED GENE TRANSFER; *AGROBACTERIUM RHIZOGENES; AGROBACTERIUM TUMEFACIENS.*

**genome:** The total genetic content of an organism, including the nuclear DNA and organellar DNA in eukaryotes or plasmid DNA in prokaryotes (Russell, 2002).

**genomic library:** A large set of random DNA fragments of the genome of an organism contained in vectors such as plasmids and cloned in a suitable host (Russell, 2002).

**genomics:** The application of sequencing, mapping, and bioinformatics to the analysis of genomes. The aim is to find genes within genomic sequences, to align sequences in databases and determine the degree of matching, to predict the structure and function of gene products, and to use sequence information in identification and in determining the relationship between organisms (Russell, 2002). *See* BIOINFORMATICS.

**genotype:** The genetic composition of an organism, which on interacting with the environment gives the phenotype (Grant, 1975).

**gentamycin:** An aminoglycoside antibiotic isolated from *Micromonospora* that is effective against *Pseudomonas* spp. (Walsh, 2003). *See* ANTIBIOTICS.

**germination:** The physiological changes that a reproductive structure such as a seed undergoes before visible signs of growth. It is influenced by availability of water, temperature, and in some cases light, and by dormancy factors (Hartmann et al., 2001). *See* DORMANCY.

**germplasm:** Syns. germ pool, germplasm pool; the genetic material of an organism. The term can be used to refer to the material in the gene pool (Herzberg et al., 1995). *See* CROP GENE POOL.

**germplasm conservation:** Large collections of crop varieties and related species are maintained in national and international research centers as living collections, seed collections, and cryopreserved tissue cultures. Ecosystems are maintained in situ in national parks and in botanic gardens, and collections of lesser crops and ornamental plants are conserved by plant breeders and amateurs (Herzberg et al., 1995; Cassells, 2002). *See* CRYOPRESERVATION.

**germplasm storage:** *See* CRYOPRESERVATION; GERMPLASM CONSERVATION.

**GFP:** *See* GREEN FLUORESCENT PROTEIN.

**gibberellic acid:** Identical to $GA_3$, gibberellic acid is a natural plant growth regulator first isolated from the fungus *Gibberella fujikuroi* (Taiz and Zeigler, 2002). $GA_3$ is used in the management of fruit setting in fruit crops and to elongate sugarcane (Basra, 2000). It is also used in tissue culture media (George, 1993; Gaba, 2005). *See* GIBBERELLIN INHIBITORS; PLANT GROWTH REGULATOR.

**gibberellin inhibitors:** Gibberellin synthesis is inhibited by ancymidol and paclobutrazol, which are used in the field (Basra, 2000) and in tissue culture, inter alia, to reduce stem elongation and induce tuberization (George, 1993).

**gibberellins:** A large group of terpenoid plant growth regulators related to $GA_3$ and associated with stem elongation in rosette plants, seed germination, breaking of dormancy and other processes

(Srivastava, 2002). *See* GIBBERELLIC ACID; PLANT GROWTH REGULATOR.

**Giemsa:** A stain used for chromosome banding. Heat or alkali denaturing of mitotic chromosomes followed by reassociation of the heterochromatic regions under controlled conditions permits the heterochromatic regions, though not the euchromatic regions, to be stained with Giemsa. The bands are specific for particular autosomal chromosomes and so aid in chromosome identification and karyotyping (Dyer, 1979). *See* KARYOTYPE ANALYSIS.

**glass bead sterilizer:** An instrument for sterilizing small items of equipment for use in tissue culture. The instruments are placed in a chamber containing glass beads heated to 220°C. *See* INSTRUMENT STERILIZATION.

**glasshouse:** *See* GREENHOUSE.

**glassiness:** *See* HYPERHYDRICITY.

**glassy shoots:** *See* HYPERHYDRICITY.

**globular embryo:** *See* EMBRYOGENESIS.

**glovebox:** A simple cabinet for tissue culture, typically made from a white formica base, back, sides, and front and a sloped Perspex top. The front panel has two round hand holes (George, 1993).

**glucose:** An aldohexose sugar. Glucose is the main sugar used for energy generation in the respiration of most organisms (Metzler, 2001). It is the basic unit of starch and cellulose and with fructose forms sucrose, the common energy source in plant tissue culture media (George, 1993; Beyl, 2005).

**glutamic acid:** One of the 20 amino acids commonly found in proteins. Glutamic acid has an acidic carboxyl group on its side chain that can serve as both an acceptor and a donor of ammonia. Glutamic acid can also be converted reversibly to α-ketoglutaric acid, an intermediate in the citric acid cycle (Metzler, 2001). *See* ASPARTIC ACID; GLUTAMINE.

**glutamine:** An amino acid derived from the reaction of ammonia with glutamic acid. Glutamine can also be used by cells, like glucose, for metabolic energy (Metzler, 2001).

**glutaraldehyde:** A sterilizing agent and tissue fixative (Gahan, 1984).

**glutathione:** A sulfhydryl ($-$SH) antioxidant and enzyme cofactor. Glutathione is ubiquitous in plants and microorganisms, is water soluble, and is found mainly in the cytosol and other aqueous phases of the cell. It is one of the most concentrated antioxidants, reaching millimolar levels, and exists in two forms: "reduced glutathione" tripeptide or GSH, the antioxidant form, and the oxidized form, which is a sulfur–sulfur-linked compound (glutathione disulfide or GSSG). The GSSG/GSH ratio is used as an indicator of oxidative stress (Inze and van Montague, 2001; Gaspar et al., 2002).

**glycine:** One of the 20 amino acids commonly found in proteins. Glycine is involved in the biosynthesis of heme (a component of chlorophyll), DNA purine bases, and glutathione (Metzler, 2001).

**glycine betaine:** An alkaloid and osmolyte associated with abiotic stress responses (Basra and Basra, 1997; Taiz and Zeigler, 2002). *See* ABIOTIC STRESS; COMPATIBLE SOLUTES.

**glyphosate:** An organophosphorus herbicide. *See* GENETIC ENGINEERING; ROUNDUP.

**GMOs:** *See* GENETICALLY MODIFIED ORGANISMS.

**good laboratory practice:** Based on carrying out and implementing an HACCP (hazard analysis critical control points) analysis of the laboratory (Mortimore and Wallace, 2001). This involves the identification of the critical control points in the process and implementing a strategy to control risks. Key elements are laboratory design, staff training, equipment maintenance, sound protocols, and monitoring for indicator microorganisms and contamination (Leifert and Cassells, 2001; Cassells and Doyle, 2004; Figure 13). *See* HAZARD ANALYSIS CRITICAL CONTROL POINTS; MICROPROPAGATION LABORATORY.

**grafting:** A method of vegetative propagation in which a bud containing part of one plant (the scion) is inserted into another plant (the stock). The stock is usually disbudded; the scion establishes vascular connection (graft union) with the stock and grows into a stem. This method of propagation is commonly used for propagating woody plants (Hartmann et al., 2001). It is also used for "rejuvenation" of woody species in vitro (George, 1993). *See* IN VITRO REJUVENATION.

**gram stain:** A bacteriological stain to distinguish between two physiologically different groups of bacteria: the gram-positive and gram-negative bacteria. The bacteria are stained with a basic dye, e.g., crystal violet, which is bound to the material with iodine (a mordant). The dye complex cannot be removed from gram-positive bacteria by destaining with acetone or alcohol, whereas it is washed out of gram-negative bacteria. On counter- (second) staining with carbol fuchsin, gram-positive bacteria appear violet, whereas gram-negative bacteria take up the red counter stain (Lelliot and Stead, 1987; Prescott et al., 2001; Schaad et al., 2001).

**green fluorescent protein:** GFP; a protein that emits green fluorescence when illuminated by blue light, used as a marker for transformed cells in a plant. This presents an alternative to the *GUS* and luciferase assays (Gray, Jayasankar, et al., 2005). *See* β-GLUCURONIDASE GENE; GENETIC ENGINEERING.

**greenhouse:** A structure of glass or plastic for the protection of plants, mainly against cold stresses (Hartmann et al., 2001). *See* MICROPLANT ESTABLISHMENT.

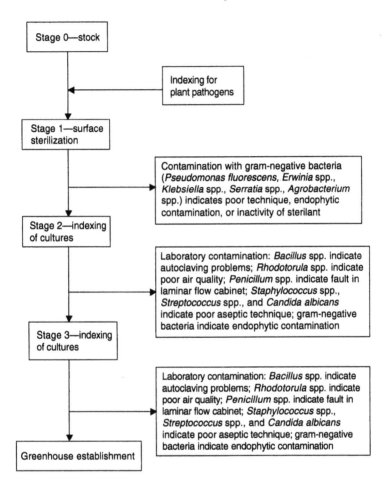

FIGURE 13. A flow diagram illustrating the principle of plant pathogen and bacterial contamination management in tissue culture (from Cassells and Doyle, 2004). Stock material is indexed for plant pathogens (see Table 2). If diseased, stock plants may be treated (*see* THERMOTHERAPY; CHEMOTHERAPY), or meristem or *in vitro* thermo- and chemo-therapy can be applied. Laboratory contamination may be controlled with good working practice. The diagram shows indicator organisms which can be used to monitor sources of contamination at each stage in micropropagation (Cassells and Doyle, 2004, 2005). *See* GOOD LABORATORY PRACTICE; HAZARD ANALYSIS CRITICAL CONTROL POINTS; PATHOGEN ELIMINATION.

**green islands:** Areas of callus in which chlorophyll develops. Green islands are associated with competence for adventitious regeneration (Cassells, 1979; Curry and Cassells, 1998).

**growth regulators:** *See* PLANT GROWTH REGULATOR.

**growth retardants:** Chemicals that inhibit plant growth. Both abscisic acid and ethylene inhibit growth at low concentration; antigibberellins inhibit stem elongation (Basra, 2000). *See* PLANT GROWTH REGULATOR.

**growth room lighting:** The artificial lighting provided for plant growth in vitro. Growth room photosynthetically active radiation (PAR) lighting is usually provided by daylight fluorescent tubes in the range 30-50 $\mu$mole·m$^{-1}$·sec$^{-1}$. Some consider that supplementary red light from tungsten or other light sources is beneficial (George, 1993). Microplants in light intensities in excess of a photosynthetic photon flux (PFF) of 100 $\mu$mole·m$^{-1}$·sec$^{-1}$ may benefit from increased carbon dioxide supply (Kozai, 1991; Zobayad et al., 2000). *See* ARTIFICIAL LIGHT.

**growth rooms:** Rooms in which the temperature is controlled to ±1-2°C. Growth rooms are provided with artificial photosynthetically active radiation (PAR) light usually with a photon fluorescent flux (PFF) of 30-100 $\mu$mole·m$^{-2}$·sec$^{-1}$. Day length is also controlled. Humidity may be controlled (George, 1993). *See* ARTIFICIAL LIGHT; AUTOTROPHIC CULTURE; PHYTOTRON.

**growth stages:** Following inoculation: lag phase, exponential growth phase, linear growth phase, phase of decreasing growth, stationary phase, and culture inviability (Figure 14). *See* CELL SUSPENSION CULTURE.

**growth substances:** *See* PLANT GROWTH REGULATOR.

**guard cells:** A pair of specialized cells surrounding the stomatal pore (Esau, 1977; Taiz and Zeigler, 2002). Malfunction of the stomata is

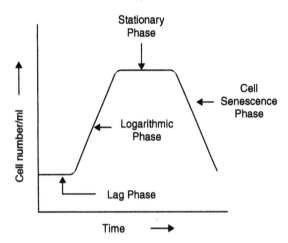

FIGURE 14. Growth stages in a cell suspension culture. *See* CELL CULTURE.

associated with failure of microplants at establishment (Ziv and Ariel, 1994). *See* MICROPLANT ESTABLISHMENT.

*GUS gene: See* β-GLUCURONIDASE GENE.

**gymnosperms:** A class of seed-bearing plants differing from the angiosperms in that the seeds lack an enclosing carpellary structure ("naked" seeds). Gymnosperms include the commercially important conifers (Hartmann et al., 2001; Graham et al., 2003).

**H** **habit:** Form taken by a plant.

**habituated:** Accustomed to a new environment. Some long-term callus cultures become independent of exogenous auxin and cytokinin by synthesizing their own (Binns and Meins, 1979; George, 1993).

**HACCP:** *See* HAZARD ANALYSIS CRITICAL CONTROL POINTS.

**haemocytometer:** *See* HEMOCYTOMETER.

**hair:** Single or multiple cell structure for absorbance or secretion. Unlike the more glandular trichomes, hair is epidermal in derivation (Esau, 1977).

**hairy roots:** *See AGROBACTERIUM RHIZOGENES.*

**halophyte:** Plant adapted to a high-salt environment (Basra and Basra, 1997; Taiz and Zeigler, 2002).

**hanging droplet:** Droplet of culture medium hanging from the underside of a coverslip over a well slide in which cells or protoplasts can be cultured. Hanging droplets were originally employed as a means of testing a range of media component concentrations using small volumes and small amounts of tissue (George, 1993).

**haploid:** Plants with one set of chromosomes. In the case of polyploid plants, plants with half the number of chromosomes as the parent (Grant, 1975). *See* ANTHER; POLLEN CULTURE.

**haploid induction:** *See* ANTHER; POLLEN CULTURE.

**haplontic:** Life cycle in which the haploid phase predominates and the diploid phase is limited to the zygote.

**hardening-off:** *See* WEANING OF PLANTS.

doi:10.1300/5648_08

**hazard analysis critical control points:** HACCP; a management system in which food safety is addressed through the analysis and control of biological, chemical, and physical hazards from raw material production, procurement, and handling, through to manufacturing, distribution, and consumption of the finished product (Mortimore and Wallace, 2001). The principles of HACCP can be applied in the tissue culture laboratory to control microbial contamination (Leifert and Cassells, 2001; Cassells and Doyle, 2004). *See* GOOD LABORATORY PRACTICE.

**health certification:** *See* PLANT HEALTH CERTIFICATION.

**health status:** Determined by carrying out appropriate tests for the presence of plant pests and pathogens (Fox, 1993). *See* PLANT HEALTH CERTIFICATION.

**heart-shaped embryos:** *See* EMBRYOGENESIS.

**heat labile:** Broken down on heating, as some components in culture media are (George, 1993; Beyl, 2005). *See* FILTER-STERILIZATION.

**heat pump:** Used for temperature control in laboratories and greenhouses.

**heat shock:** Abiotic stress imposed by excessive heat (Taiz and Zeigler, 2002). *See* ABIOTIC STRESS; BIOTIC STRESS.

**heat treatment:** *See* THERMOTHERAPY.

**Heller medium:** Mineral salt formulations for plant tissue culture Heller (1953, 1955).

**hematoxylin:** A dye extracted from *Haematoxylon campechianum,* which on oxidation to haematin is used to stain nuclei and cellulose blue in histology. Hematoxylin is used in cytochemistry to react with phospholipids (Gahan,1984).

**hemicellulase:** A hydrolytic enzyme breaking down hemicellulose to galactose. It is used in the preparation of protoplasts (Warren, 1991).

**hemicellulose:** Plant cell wall polysaccharide, which together with pectin and lignin forms the cell wall matrix (Brett and Waldron, 1989; Smith, 1993; Carpita et al., 2001). *See* PLANT CELL WALL.

**hemiendophyte:** A localized contamination of plant tissues, usually caused by an epiphyte (plant-surface-inhabiting bacterium). Such contaminants may escape explant surface sterilization (Cassells and Doyle, 2004). *See* ENDOPHYTIC ORGANISMS; EPIPHYTE; SURFACE STERILIZATION.

**hemizygous:** Describes a single gene copy with no allele, as in haploid plants (Allard, 1999).

**hemocytometer:** A chamber for counting blood cells under a microscope; also used to count plant cells and protoplasts.

**HEPA filter:** High-efficiency particulate air filter; used in laminar flow cabinets for removing particles larger than 0.3 μm (George, 1993).

**herbaceous plants:** Plants used in medicine, cooking, and perfume making. They are nonwoody perennial plants that die back at the end of the season (Heywood, 1993; Hartmann et al., 2001).

**herbicide:** A chemical whose application causes a plant to either die or have severely inhibited growth; weedkiller (Basra, 2000; Cobb and Kirkwood, 2000).

**herbicide resistance:** Partial or complete resistance to herbicides (Roe et al., 1997). *See* GENETIC ENGINEERING.

**hereditary:** Describes characteristics transferred genetically between subsequent generations (Grant, 1975).

**heteroblasty:** Progressive change in form and complexity of successive organs on moving from juvenile to adult forms, e.g., the leaves of ivy (Howell, 1998). *See* HETEROPHYLLY; PHASE CHANGE.

**heterochromatin:** Appears as heavily stained clumps of chromatin in the interphase nucleus. Heterochromatin is the site of very few genes, which are silenced, and is comprised primarily of repetitive sequences. Recombination rates are low, and reporter genes present in heterochromatin are transcriptionally repressed. The heterochromatic gene silencing is passed to daughter cells. Heterochromatin is probably stabilized by the hyperacetylation of histones H3 and H4 and silenced by methylation and RNAi; it is especially associated with the centromere and the telomeres. Up to 20 times as many transposons are present in heterochromatin as in euchromatin. Some researchers distinguish between "constitutive heterochromatin," which contains the repetitve sequences, and "facultative heterochromatin," which contains those genes temporarily silenced (Turner, 2001; Martienssen et al., 2003).

**heterogeneous:** Nonuniform, e.g., cells in a tissue or callus.

**heterograft:** Interspecific graft (Hartmann et al., 2001).

**heterokaryon:** Multinucleate product of fusion of unlike cells with different nuclei (Warren, 1991). *See* PROTOPLAST FUSION.

**heteromixotrophic growth:** *See* MIXOTROPHIC CULTURES.

**heterophylly:** Having two or more different leaf types, which may be associated with an adaptation, e.g., to environmental conditions such as submersion (Howell, 1998).

**heterosis:** More vigorous growth, productivity, or disease resistance shown by a plant when compared with either parent (Allard, 1999).

**heterotrophic growth:** Growth dependent on a supplied carbon source. *See* AUTOTROPHIC CULTURE.

**heterozygous:** Describes an individual having different alleles of a single gene or from gametes with a different arrangement of genes, e.g., due to inversion. Heterozygosity is inferred when an individual does not breed true for a trait (Allard, 1999). *See* HOMOZYGOUS.

**hexose monophosphate shunt:** Shunt metabolic pathway from glucose-6-phosphate in glycolysis and back to either fructose-6-phosphate or glyceraldehyde-3-phosphate; after decarboxylation from a 6- to a 5-carbon sugar no further energy is required for onward transformations to 3-, 4-, 5-, 6-, and 7-carbon sugars. Most steps do not require energy, and the $NADPH_2$ (reduced nicotinamide adenine dinucleotide phosphate) produced is used in lipid biosynthesis (Smith, 1993; Bryant et al., 1999; Metzler, 2001).

**hexose sugar:** Six-carbon sugar molecule, e.g., glucose, fructose (Smith, 1993; Bryant et al., 1999).

**high-performance liquid chromatography:** HPLC; a technique for separating molecules based on the migration coefficients in a solvent/matrix combination (McMaster, 1994).

**histochemistry:** The quantitative study of the chemistry and metabolism of cells and tissues using chemical color reactions (Gahan, 1984).

**histogen theory:** Concept of tissue formation in which the meristem is considered to consist of three zones: the dermatogen, giving rise to the epidermis; the periblem, to the cortex; and the plerome, to all primary tissues within the cortex (Howell, 1998). *See* APICAL MERISTEM.

**histology:** Microscope study of the structure of tissues.

**histone:** Common histones include H1, H2, H2A, H2B, H3, and H4, which combine with DNA to form nucleosomes as part of eukaryotic chromosome structure. Histone is absent from prokaryotes (Metzler, 2001).

**histone acetylation:** Strongly influences internucleosomal interactions. High acetylation constrains wrapping of DNA on the nucleosome surface; weak acetylation does not constrain DNA, which is accessible to transcription factors. Histone acetylation may play a role in epigenetic phenomena (Russell, 2002). *See* DNA METHYLATION; EPIGENETICS.

**Hoagland's solution:** A mineral nutrient formulation that forms the basis for many hydroponic and autotrophic tissue culture media (Hoagland and Arnon, 1938; Hoagland, 1948).

**homohistont:** A nonchimeric plant (Tilney-Bassett, 1991).

**homozygous:** Individual formed from gametes either bearing identical alleles of a specific gene or resembling each other in gene arrangement. Homozygous individuals breed true for a given character (Allard, 1999).

**hood:** *See* FUME HOOD.

**hormone:** Animal-derived concept: a molecule that is produced at one site, enters directly into the bloodstream, and affects a specific event at another site. The concept is difficult to apply specifically to plants, although the term is used to loosely describe naturally occurring plant growth regulators or plant bioregulators (Taiz and Zeigler, 2002). *See* PLANT GROWTH REGULATOR.

**HPLC:** *See* HIGH-PERFORMANCE LIQUID CHROMATOGRAPHY.

**humidity:** *See* RELATIVE HUMIDITY.

**hybrid:** $F_1$ heterozygotic progeny of two individuals differing in one or more genes (Allard, 1999).

**hybridization:** Production of a hybrid; the annealing of either DNA/DNA or DNA/RNA strands from two organisms to determine complementarity of genes at annealing sites (Russell, 2002). *See* INCOMPATIBILITY; PROTOPLAST FUSION.

**hydrochloric acid:** A strong acid, usually at a maximal concentration of 12.6 N.

**hydrogel:** A commercial gelling agent. *See* GELLING AGENT.

**hydrogen peroxide:** $H_2O_2$; oxidizing agent that on dilution to ~3 percent is used to surface-sterilize plant material; a mutagen. Hydrogen peroxide is a stress-signaling compound in plants (Inze and van Montague, 2001). *See* OXIDATIVE STRESS.

**hydrolase:** Enzyme catalyzing hydrolytic reactions. Hydrolases include phosphatases, esterases, deoxyribonucleases, ribonucleases, lipases, proteases, and carbohydrases (Metzler, 2003).

**hydroponics:** Growth of plants in a liquid culture medium. This technology is widely used in horticulture and is also referred to as soilless culture (Sholto-Douglas, 1976; Resh, 2002). *See* AEROPONICS; AUTOTROPHIC CULTURE.

**hydroxyproline:** An amino acid produced by hydroxylation of proline. With proline, it is one of two cyclic amino acids found in proteins (Metzler, 2003). *See* AMINO ACIDS.

**8-hydroxyquinoline:** An antibiotic effective against some fungi and bacteria. It is a component of photographic developer. *See* ANTIBIOTICS; ANTIMYCOTICS.

**hygromycin B:** An aminoglycosidic antibiotic produced by *Streptomyces hygroscopicus* (Walsh, 2003). *See* ANTIBIOTICS.

**hygromycin phosphotransferase:** A marker system for plant transformation. It inactivates the antibiotic hygromycin B. It is one of the preferred antibiotic resistance marker systems for transformation of monocot plants, particularly Gramineae (cereals and forages) (Slater et al., 2003). *See* GENETIC ENGINEERING.

**hygroscopic:** Able to take up water from the environment.

**hyperhydricity:** Syn. vitrification (see Debergh et al., 1992); a syndrome reflecting adverse factors in vitro. The signs are glassy hyperhydrated leaves, shoot and leaf tip necrosis, basal callusing, and loss of apical dominance in the cultures. Many components of the medium and culture vessel atmosphere have been implicated in this phenomenon (Ziv, 1991; Cassells and Roche, 1994; Cassells and Walsh, 1994; Ziv and Ariel, 1994; Kevers et al., 2004).

**hypermethylation:** High DNA methylation associated with gene silencing and aging of cells and tissues (Matzke and Matzke, 2000). *See* DNA METHYLATION; EPIGENETICS; HISTONE ACETYLATION; RNA INTERFERENCE.

**hyperplasia:** Abnormal overdevelopment with increased number of cells, e.g., galls (Agrios, 1997).

**hyperploid:** Cell or organism with an extra chromosome or chromosome fragment.

**hypersensitivity:** Disease-resistance mechanism in which plant response involves the death of cells adjacent to an infective agent and induction of systemic acquired resistance (Agrios, 1997; Strange, 2003). *See* BIOTIC STRESS; SYSTEMIC ACQUIRED RESISTANCE.

**hypertonic:** Describes a solution having higher osmotic potential than adjacent cells, resulting in plasmolysis of the cells (Ridge, 2002). *See* PROTOPLAST ISOLATION.

**hypertrophy:** Abnormal increase in cell size leading to abnormal increase in tissue size (Agrios, 1997; Strange, 2003).

**hypha:** Branched filament, many of which make up a fungal mycelium (Watkinson and Gooday, 2001).

**hypochlorite:** The salt of hypochlorous acid, yielding an oxidizing agent, e.g., sodium hypochlorite, used to surface-sterilize plant material (George, 1993; Beyl, 2005). *See* SURFACE STERILANTS.

**hypocotyl:** Embryonic zone between the cotyledons and the radicle; the transition zone between the vascular bundle arrangement of the stem and that of the root (Esau, 1977).

**hypomethylation:** Low DNA methylation associated with young tissues (Russell, 2002; Joyce et al., 2003). *See* DNA METHYLATION; EPIGENETICS.

**hypoplastism:** Underdevelopment of plants leading to stunting and dwarfism through the effect of either disease or lack of nutrients (Agrios, 1997).

**hypoploid:** Cell or organism lacking one or more chromosomes or chromosome fragments than the euploid number (Dyer, 1979). *See* ANEUPLOID.

**hypotonic:** Describes a solution having lower osmotic potential than an adjacent solution, e.g., cell sap, resulting in increased turgidity of cells (Ridge, 2002).

**hypoxia:** Deficiency of oxygen in the tissues associated with waterlogging (Taiz and Zeigler, 2002). *See* ABIOTIC STRESS.

**IAA:** *See* INDOLEACETIC ACID.

**IBA:** *See* INDOLEBUTYRIC ACID.

**ice crystals:** *See* CRYOPRESERVATION.

**image analysis:** The digitization of images, which can then be analyzed using computer software programs (Gonzalez and Woods, 2001). The information gained, especially from color imaging, can be used in the identification of species and varieties, in studies on ontogeny, and in quantifying disease in plants (Cazzulino et al., 1991; Cassells et al., 1997, 1999). Image analysis is used in microscopy to analyze a number of aspects of cell and tissue structure and function. *See* THERMAL IMAGING.

**immerse:** To push below the surface of a liquid. *See* TEMPORARY IMMERSION.

**immobilized cells:** *See* CELL IMMOBILIZATION.

**immunization:** The induction of antibodies by injection of a foreign antigen, commonly into a rabbit or larger mammal (Roitt et al., 2001; Coico et al., 2003). *See* ANTIBODY; ANTIGEN.

**immunoelectron microscopy:** *See* IMMUNOSORBENT ELECTRON MICROSCOPY.

**immunofluorescence:** The labeling of antibodies with a fluorescent dye (e.g., FITC) for detection by fluorescence microscopy (Murphy, 2001). *See* FLUORESCENCE MICROSCOPY.

**immunosorbent electron microscopy:** ISEM; the use of antibodies to detect viral and other antigens in electron microscopy. The antibodies may be coated on the electron microscope grid to increase the sensitivity of virus detection or applied after virus particle adhesion to decorate the particles; i.e., the antibody specifically binds to target

doi:10.1300/5648_09

particles and may be used to identify specific viruses or virus strains in a mixed preparation. The antibodies may also be linked to gold particles to aid localization of antigens in transmission electron microscope preparations (Hari and Das, 1998). *See* VIRUS COMPLEX.

**import license:** An official document required under international phytosanitary legislation for the importation of plant material from another jurisdiction. Local restrictions on importation may apply. The import license specifies the conditions for importation, which must be satisfied by the plant health certificate accompanying the consignment. *See* PLANT HEALTH CERTIFICATION.

**inbreeding:** Breeding between close relatives; naturally self-pollinating crops are genetically equivalent to vegetative propagation.

**incompatibility:** Genetically determined inability of gametes from genetically similar plants to form fertile seed (Grant, 1975). *See* EMBRYO RESCUE; IN VITRO POLLINATION.

**incubator:** An apparatus for the cultivation at controlled temperature, with or without illumination, of microorganisms.

**indeterminate growth:** Growth that is not limited by the genetic developmental program of the plant as, e.g., in most trees (Hartmann et al., 2001). *See* DETERMINATE GROWTH.

**indexing:** Examination of plant material for the presence of contaminating microorganisms (Hadidi et al., 1998). Indexing, applied to tissue cultures, may involve testing for the presence of plant pathogens and environmental microorganisms, which may contaminate the cultures (Cassells and Doyle, 2004). *See* PATHOGEN INDEXING.

**indicator microorganisms:** Microorganisms associated with events such as biological contamination of air and water. They are used to monitor tissue culture laboratory practices and facilities (Leifert and Cassells, 2001; Cassells and Doyle, 2004, 2005). *See* GOOD LABORATORY PRACTICE.

**indicator plant:** A plant that reacts to pathogen inoculation by developing characteristic symptoms. Such plants are used in tests to identify viruses and virus strains and to confirm the pathogenicity of bacterial and fungal isolates. They are also used to confirm negative results in ELISA and other molecular diagnostic tests (Lelliot and Stead, 1987; Dijkstra and de Jeger, 1998). *See* ENZYME-LINKED IMMUNOSORBENT ASSAY; POLYMERASE CHAIN REACTION.

**indirect embryogenesis:** The formation of embryos from callus as opposed to the primary cells of the explant (George, 1993). *See* DIRECT EMBRYOGENESIS; SECONDARY EMBRYOS.

**indirect organogenesis:** The production of roots and shoots from callus. Shoots produced in this way may be genetically variable (George, 1993; Mohan Jain et al., 1998). *See* DIRECT ORGANOGENESIS; MICROPROPAGATION; SOMACLONAL VARIATION.

**indoleacetic acid:** IAA; an auxin that occurs universally in plants. *See* AUXINS; PLANT GROWTH REGULATOR.

**indolebutyric acid:** IBA; 4-indol-3-ylbutyric acid, a natural auxin. *See* PLANT GROWTH REGULATOR.

**indolepyruvic acid:** An intermediate in one of the IAA biosynthetic pathways (Taiz and Zeigler, 2002). *See* AUXINS.

**induced gene:** A gene that is not constitutively expressed but is activated by a chemical or environmental agent (Russell, 2002).

**induced resistance:** Plant biotic stress resistance induced by artificial inoculation with microorganisms or treatment with chemical inducers ("elicitors") (Strange, 2003). *See* INDUCED SYSTEMIC RESISTANCE; SYSTEMIC ACQUIRED RESISTANCE.

**induced systemic resistance:** ISR; induced plant-growth-promoting rhizobacteria (PGPR) resistance dependent on ethylene and jasmonic acid signaling. ISR can occur without the production of pathogenesis-related proteins (van Loon, 2000). *See* PLANT-GROWTH-PROMOTING RHIZOBACTERIA; SYSTEMIC ACQUIRED RESISTANCE.

**induction medium:** A medium on which cells dedifferentiate and become competent for determination (Christianson and Warnick, 1985; Schwarz et al., 2005). *See* COMPETENCE; DEDIFFERENTIATION; EXPRESSION MEDIUM; INDUCTION MEDIUM.

**indumentum:** a covering of hair or integument.

**inflorescence:** A flowering stem containing more than one flower (Heywood, 1993).

**inoculation:** The transfer of a microorganism onto a plant or of a microorganism or cell or tissue onto a medium.

**inorganic salts:** *See* MINERAL NUTRITION.

**inositol:** Myo-inositol; a cyclic hexahydric alcohol that is a component of phosopholipids, phytic acid, and phytin (Metzler, 2001). Myo-inositol is a component of plant tissue culture media (George, 1993; Beyl, 2005). *See* PLANT TISSUE CULTURE MEDIA.

**instrument sterilization:** In preparation for tissue culture, instruments are commonly sterilized in foil with indicator paper in an autoclave. When in use, the instruments are sterilized between operations by immersion in alcohol with flaming or by insertion in either a bacticinerator or a glass bead sterilizer. *See* AUTOCLAVE; AUTOCLAVE TAPE; BACTICINERATOR; GLASS BEAD STERILIZER.

**International Union for the Protection of New Varieties of Plants:** UPOV; An intergovernmental organization with headquarters in Geneva (Switzerland). It was established by the International Convention for the Protection of New Varieties of Plants, adopted in Paris in 1961 and revised in 1972, 1978, and 1991. The objective of the convention is the protection of new varieties of plants by an intellectual property right.

**internode:** The section of a stem between the nodes lacking buds. Internodes are used as explants in tissue culture. Caution: internodes may carry over pathogens into culture (Cassells and Doyle, 2004, 2005).

**interspecific hybridization:** Hybridization between different species. Interspecific hybrids are achieved by crossbreeding in vivo. Where interspecific hybrids are infertile, fertility may be restored by colchicine treatment (Allard, 1999). Interspecific hybridization may also be attempted using protoplast fusion (Taji et al., 2001; van Tuyl et al., 2002). *See* COLCHICINE; CROP GENE POOL; SOMATIC HYBRIDIZATION.

**intracuticular wax:** The wax embedded within the cutin matrix of the epidermal cells. *See* EPICUTICULAR WAX.

**inverted microscope:** A microscope in which the objectives are situated below the specimen stage and hence the specimen is viewed from below.

**in vitro:** Lit. "in glass"; includes artificial culture, i.e., tissue culture.

**in vitro atmosphere:** The gases in the tissue culture vessel, which are influenced by the permeability of the vessel's walls and the type of closure or vessel lid (Cassells and Roche, 1994). A glass vessel is impermeable, but plastics may be permeable or differentially permeable to gases, allowing, e.g., ethylene to accumulate (Cassells and Walsh, 1994). *See* PLASTIC FILMS.

**in vitro pollination:** A method in which pollen is placed beside an excised ovule on aseptic culture media to bypass incompatibility mechanisms (Taji et al., 2001). *See* PREZYGOTIC INCOMPATIBILITY.

**in vitro reinvigoration:** *See* IN VITRO REJUVENATION.

**in vitro rejuvenation:** Tissue culture used to rejuvenate buds from mature tissues to facilitate clonal propagation of elite individuals. The protocol may involve in vitro grafting to seedling rootstock or serial subculture of the microshoots. Microplants are considered to be more juvenile than cuttings and other vegetative propagules

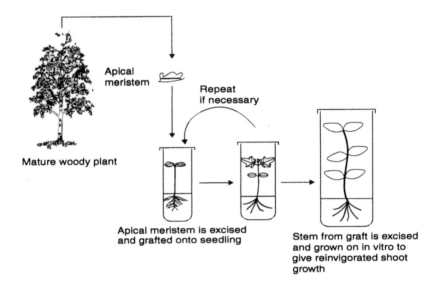

Apical
meristem

Repeat
if necessary

Mature woody plant

Apical meristem is excised
and grafted onto seedling

Stem from graft is excised
and grown on in vitro to
give reinvigorated shoot
growth

FIGURE 15. *In vitro* reinvigoration ("in vitro rejuvenation"). An apical meristem from a mature shoot is grafted to an aseptically germinated seed of the same species *in vitro*. The established meristem may be serially subcultured to new seedlings or the shoot excised and transferred to rooting medium. *See* IN VITRO REJUVENATION.

and more mature than seedlings (George, 1993; Figure 15). *See* REINVIGORATION; REJUVENATION.

**in vitro rooting:** Stage 3 of micropropagation, the induction of roots on microshoots in vitro, usually involving transfer to a reduced-strength medium containing an auxin or high auxin to low cytokinin ratio (George, 1993; Figure 16). *See* MICROPROPAGATION.

**in vitro weaning:** Microplants from high-sucrose media and from high-humidity in vitro environments may have poorly developed physiological process (Grunewaldt-Stocker, 1997), e.g., poorly developed photosynthetic apparatus, nonfunctional stomata, and thin cuticles (Preece and Sutter, 1991; Cassells and Walsh, 1994; Kiersteins, 1994). In vitro weaning involves strategies to improve the physiological quality of microplants in Stage 3 of micropropagation, prior to establishment. Strategies involve transfer to sugar-free (autotrophic) culture (Kozai, 1991) and induction of a transpiration stream in the microshoots by

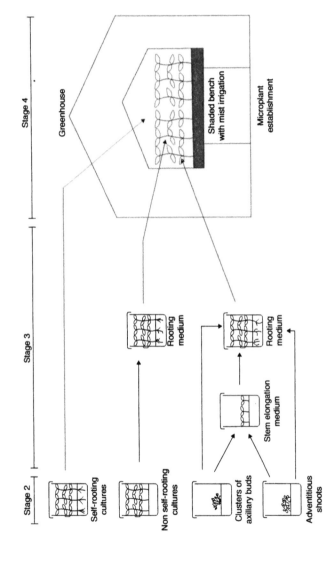

FIGURE 16. Strategies for the rooting of microshoots *in vitro* (Stage 3 of micropropagation) or *extra vitrum* (Stage 4 of micropropagation). Shoots which have spontaneously rooted ("self-rooted") in Stage 2 may be transferred directly to the greenhouse for establishment, otherwise shoots from Stage 2 cultures are transferred to a rooting medium in Stage 3. If axillary bud proliferation or adventitious regeneration is used for cloning in Stage 2 it may be beneficial to split Stage 3 into a stem elongation step (Stage 3a) followed by the rooting step (Stage 3b). *See* IN VITRO ROOTING, Exhibit 2 (Stages of micropropagation).

ventilating the culture vessels by opening the lids or using as lids plastic films permeable to water vapor (Cassells and Walsh, 1994). *See* AUTO-TROPHIC CULTURE; BIOTIZATION; MICROPLANT ESTABLISHMENT; MICROPROPAGATION; TISSUE CULTURE VESSELS.

*in vivo*: Describes biological processes occurring in the natural environment.

**iodine:** I; nonessential element for plant growth (Marschner, 1994) sometimes included in tissue culture media (George, 1993).

**ionic concentration:** *See* PLANT TISSUE CULTURE MEDIA.

**ioxynil:** A nitrile herbicide (Basra, 2000).

**2-iP:** *See* ISOPENTENYL ADENINE.

**IPA:** *See* ISOPENTENYL ADENINE.

**iron:** Fe; an essential plant macroelement necessary for the synthesis of chlorophyll, cytochromes, ferredoxin, and several enzymes (Marschner, 1994). Usually provided as an iron chelate to maintain its solubility in tissue culture media (Gautheret, 1959; George, 1993). *See* CHELATING AGENTS; CULTURE MEDIUM.

**irrigation:** Provision of water for plant growth. *See* WATER QUALITY.

**ISEM:** *See* IMMUNOSORBENT ELECTRON MICROSCOPY.

**isoelectric focusing:** An electrophoretic technique in which proteins are separated in a pH gradient where they migrate to the point at which they have no net charge (Hames and Rickwood, 2001). *See* GEL ELECTROPHORESIS; ISOELECTRIC POINT.

**isoelectric point:** The pH at which a protein carries no net charge. *See* ISOELECTRIC FOCUSING.

**isoenzymes:** *See* ISOZYMES.

**isolation medium:** A medium for the isolation of a pathogen from a plant. Generally containing a selective substrate and inhibitors of nontarget microorganisms (Lelliot and Stead, 1987; Schaad et al., 2001).

**isopentenyl adenine:** $N^6$-(2-isopentenyl)adenine; naturally occurring cytokinin. *See* CYTOKININS.

**isotonic:** Of equal osmotic potential. *See* OSMOTIC POTENTIAL; PLASMOLYSIS.

**isozymes:** Forms of enzymes that have similar specificity but may differ in pH or isoelectric point. Isozymes are used in varietal identification. *See* DNA FINGERPRINTING; ISOELECTRIC POINT.

**jasmonic acid:** Jasmonic acid and its methyl esters are widespread in plants. They are involved in growth regulation, development, leaf senescence, and defense against biotic stress. Jasmonic acid is antagonized by cytokinins (Taiz and Zeigler, 2002; Gaba, 2005). *See* PLANT GROWTH REGULATOR.

**joule:** The joule per square meter ($j \cdot m^{-2}$) is used as a unit of light measurement.

**jumping genes:** *See* TRANSPOSON.

**juvenile phase:** The immature, usually nonreproductive, form of a plant rather than the young age of a plant. The juvenile form is often phenotypically different from the adult form, e.g., leaves of *Hedera helix* (Howell, 1998). *See* ADULT PHASE.

**juvenility:** The condition of being juvenile. In the cases of woody plants, juvenility is associated with the relative ease of rooting cuttings from juvenile as opposed to mature tissues (Hartmann et al., 2001). *See* IN VITRO REJUVENATION.

doi:10.1300/5648_10

**kanamycin:** An aminoglycoside antibiotic used to select for transformed cells (Slater et al., 2003). *See* ANTIBIOTICS; GENETIC ENGINEERING; SELECTABLE MARKER.

**kanamycin resistance:** Conferred by the *nptII* gene, which is widely used as part of the construct to genetically engineer plant cells. Transformed cells can grow on kanamycin; untransformed cells are killed (Slater et al., 2003). *See* GENETIC ENGINEERING; SELECTABLE MARKER.

**Kao and Michayluk medium:** A medium originally developed for cell and protoplast culture (Kao and Michayluk, 1975).

**karyotype:** The physical appearance of stained chromosomes at mitotic metaphase (Russell, 2002). *See* CHROMOSOME.

**karyotype analysis:** Arrangement of the chromosomes in order of size and position of the centromere (Dyer, 1979).

**karyotyping:** *See* KARYOTYPE ANALYSIS.

**kinetin:** 6-Furfurylaminopurine; plant growth regulator. Kinetin is a synthetic cytokinin that promotes cell division (Taiz and Zeigler, 2002). The ratio of auxin to cytokinin and their relative concentrations are important in influencing cell division and organogenesis in cell and tissue culture (Skoog and Miller, 1957; George, 1993). *See* BENZYLADENINE; CYTOKININS; ZEATIN.

**King's medium:** A bacterial medium used for the culture of pseudomonads producing fluorescent pigment (King et al., 1954).

**L**

**l:** *See* LITER.

**labeling:** Tagging of a molecule with a radioactive atom or fluorescent molecule. *See* BIOTIN LABELING; RADIOLABELING.

**lag phase:** Phase immediately after inoculating a culture during which cell number does not increase though much physiological activity takes place (George, 1993). *See* GROWTH STAGES.

**lamina:** Flattened, bladelike part of a leaf attached to the petiole.

**laminar flow cabinet:** Enclosed work area with uniform flow of filtered air to provide a sterile environment (George, 1993; Figure 17). *See* MICROPROPAGATION LABORATORY.

**lanolin paste:** A fatty material from wool used as a base for applying hormones to plants.

**latent contamination:** Contamination that is not visible to the eye. Such contamination may be reflected in a halolike or opaque zone around the explant. Tissue cultures should be routinely indexed for contaminants as a precaution against the clonal transmission of such contaminants (Cassells and Doyle, 2004, 2005). *See* CULTURE INDEXING; GELLING AGENT.

**latent infection:** Stage of pathogen or pest infection without visible signs (Cassells and Doyle, 2004, 2005). *See* PATHOGEN INDEXING.

**lateral:** One side of an organism, as opposed to a terminal. *See* TERMINAL.

**lateral bud:** Axillary bud, the bud in the axil of a leaf (Trigiano and Gray, 2005). *See* MICROPROPAGATION.

doi:10.1300/5648_12

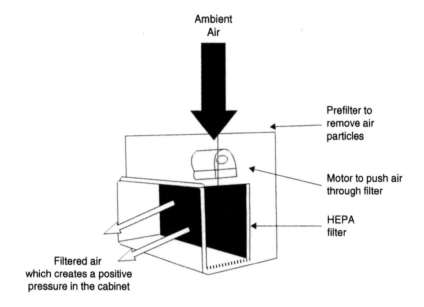

Ambient
Air

Prefilter to
remove air
particles

Motor to push air
through filter

HEPA
filter

Filtered air
which creates a positive
pressure in the cabinet

FIGURE 17. Diagram showing a laminar flow cabinet. *See* HEPA FILTER; LAMINAR FLOW CABINET; SURFACE STERILIZATION.

**lateral meristem:** Meristem positioned longitudinally in an organ, e.g., cambium. *See* APICAL MERISTEM.

**lateral root:** Root arising from the pericycle of another root (Anderson et al., 1997).

**lateral shoot:** A shoot arising from a lateral bud. *See* NODAL CULTURE.

**latex:** Fluid produced in lactiferous ducts in higher plants. Latex contains various substances either in solution or in suspension, e.g., alkaloids, rubber. It may bind or inhibit microorganisms, so preventing their detection.

**latex agglutination:** A popular, sensitive method for the detection of viruses and other antigens based on the clumping of antibody-coated latex particles in the presence of the antigen.

**latin square:** Analogous to a random block design in two directions, where the number of treatments is the same as the number of replicates (Dijkstra and de Jager, 1998; Compton, 2005). *See* RANDOMIZED BLOCK DESIGN.

**layering:** Pegging of runners and stolons to soil to promote rooting and the production of new plants (Hartmann et al., 2001); in vitro, placing of agar on cultured roots or nodal segments to encourage axillary bud proliferation (George, 1993).

**$LD_{50}$:** The dose at which 50 percent of the test organisms die, a measure of the toxicity of chemicals.

**LDPE:** *See* LOW-DENSITY POLYETHYLENE.

**leaf:** Main photosynthetic organ of a plant comprised of petiole, lamina, and leaf base (Esau, 1997; Taiz and Zeigler, 2002; Graham et al., 2003).

**leaf abscission:** *See* ABSCISSION.

**leaf area index:** LAI; ratio of total leaf surface area to the ground area available to that plant.

**leaf area ratio:** Total leaf area divided by the leaf dry weight.

**leaf buttress:** Prominent lateral site on apical meristem developing into a leaf (Esau, 1977).

**leaf explant:** A piece of tissue cut from leaves and used to establish cultures (George, 1993). *See* ADVENTITIOUS REGENERATION; MICROPROPAGATION.

**leaf mineral analysis:** Assay of the mineral content of leaves. The analysis can be used to optimize mineral nutrients for plants in culture. *See* MINERAL NUTRITION.

**leaf position:** The position on the stem. Leaves show basipetal differences in morphology and ontology (Howell, 1998). Leaf ontogeny may be related to protoplast survival potential (Cassells and Barlass, 1978). *See* PHASE CHANGE.

**leaf primordium:** Subsequent stage to leaf buttress in leaf formation (Esau, 1977).

**leaf senescence:** The period between the maturation and death of a leaf (Taiz and Zeigler, 2002).

**Leifert and Waites medium:** A medium for screening tissue cultures for bacterial contamination (Leifert and Waites, 1992).

**lethal gene:** Mutant allele resulting in the death of the organism (Russell, 2002).

**leucoplast:** Colorless plastid present in cells of roots, underground stems, and storage organs, e.g., the amyloplast, which stores starch (Smith, 1993).

**light:** Electromagnetic radiation with a wavelength range of 400 nm (violet) to 770 nm (red). *See* ELECTROMAGNETIC SPECTRUM; PHOTO-SYNTHETIC PIGMENTS.

**light flux density:** *See* PHOTON FLUX DENSITY.

**lighting:** *See* ARTIFICIAL LIGHT; GROWTH ROOM LIGHTING.

**light measurement:** *See* PAR METER.

**light microscope:** Optical microscope. A standard compound light microscope can resolve to ~0.24 μm the distance between two point sources. Owing to the limits imposed on lens construction, resolution can be increased only by illuminating with ultraviolet light of a shorter wavelength; fluorescence microscopy increases sensitivity but not resolution (Gahan, 1984; Murphy, 2001). *See* FLUORES-CENCE MICROSCOPY.

**light plants:** Plants whose photosynthesis is tolerant of high ambient light intensities and whose morphology does not change with increasing light intensity (Hartmann et al., 2001). *See* SHADE PLANTS.

**lignin:** A phenolic-derived compound found in the cell walls of xylem vessels, tracheids, and sclerenchyma and at very low levels in some parenchymal cell walls. Lignin gives strength and rigidity (Brett and Waldron, 1989; Carpita et al., 2001) and is induced in pathogen stress responses (Strange, 2003).

**limiting factor:** Environmental factor that by its variable presence or absence will control the behavior of an organism or system.

**line:** A genotype selected by a plant breeder (Allard, 1999).

**lineage:** Line of common descent from either an organism or a single cell.

**linear phase:** Exponential growth phase following the lag phase of a culture (George, 1993). *See* GROWTH STAGES.

**linkage:** Tendency for genes on the same chromosome to segregate together during meiosis (Allard, 1999).

**linkage map:** Diagrammatic representation of the relative positions of known genes on a given chromosome (Russell, 2002).

**Linsmaier and Skoog medium:** Simplified version of Murashige and Skoog (1962) medium (Linsmaier and Skoog, 1965). *See* MURASHIGE AND SKOOG MEDIUM.

**lipids:** A heterogeneous group of compounds that are either water insoluble or sparingly soluble in water but are soluble in nonpolar solvents. Lipids have either pronounced nonpolar groups or both polar and nonpolar groups and are thus either hydrophobic or amphipathic. Simple lipids are esters of either alcohols or fatty acids; complex lipids comprise a lipid plus a nonlipid component; derived lipids are, e.g., water-soluble vitamins, isoprenoids, and steroids (Murphy, 1993; Metzler, 2001).

**lipopolysaccharide:** A molecule comprised of a lipid and a polysaccharide component (Buchanan et al., 2002).

**lipoprotein:** A molecule of lipid and protein components, often linked through hydrophobic interactions (Buchanan et al., 2002).

**liposome:** Synthetic phospholipid vesicle used to transport molecules (e.g., DNA, drugs) into targeted cells. A liposome can be used to deliver material across the plasmalemma and into the cytoplasm of protoplasts (Cassells, 1978).

**liquid culture:** A stationary or a shaken culture of plant cells or tissues, in or on a culture medium or supported over a culture medium (George, 1993; Hvoslef-Eide and Preil, 2004). *See* TEMPORARY IMMERSION; TISSUE SUPPORTS.

**liquid medium:** Culture medium without added gelling agents.

**liquid nitrogen:** Condensed nitrogen gas used as a freezing agent because of its low temperature of $-196°C$. Liquid nitrogen is a poor conductor of heat, and so material is frozen in another agent cooled by liquid nitrogen. It is used in cryopreservation (Cassells, 2002). *See* CRYOPRESERVATION.

**liter:** l or L; one thousand milliliters; one decimeter cubed.

**litmus:** pH indicator, often used as litmus paper for a pH range of 4.5-8.3.

**living collections:** *See* GENE BANKS.

**Lloyd and McCown medium:** A commonly used formulation for the culture of woody plants (Lloyd and McCown, 1981).

**local lesion:** A localized spot that develops on a leaf following inoculation with a virus and is used to detect and quantify infectious virus (Dijkstra and de Jeger, 1998). Local lesions are also induced by other necrotizing pathogens, can be induced by chemical agents, and are associated with induced resistance (Strange, 2003). *See* INDICATOR PLANT; SYSTEMIC ACQUIRED RESISTANCE.

**locule:** Chamber in which specialized organs may develop (Esau, 1977).

**logarithmic phase:** Exponential or maximum growth phase of cells or microorganisms in culture, following a geometric progression (George, 1993). *See* GROWTH STAGES.

**long-day plant:** A plant requiring daily cycles with short dark periods to induce flowering (Hartmann et al., 2001).

**long-term storage:** *See* CRYOPRESERVATION; GERMPLASM CONSERVATION.

**low-density polyethylene:** LDPE; a translucent plastic with relatively high permeability to oxygen and carbon dioxide and relatively low permeabilty to water vapor (Cassells and Roche, 1994).

**low-temperature growth:** Growth of tissue culture at low temperature used for germplasm storage, sometimes combined with osmotic stress conditions (Cassells, 2002). *See* GERMPLASM CONSERVATION.

**lux:** 0.0929 foot candles (an obsolete unit for measurement of light).

**lyophilize:** *See* FREEZE-DRYING.

**lysigeny:** Destruction of cells to create a space within tissues.

**lysis:** Splitting open of cells by, e.g., enzyme hydrolysis, sonication.

**lysosome:** Single-membrane-bounded organelle containing acid hydrolases and forming part of the endomembrane system of the cell. In plants this is often the vacuole (Leigh and Sanders, 1998; De Deepesh, 2000; Alberts et al., 2002). *See* VACUOLE.

**MAb:** *See* MONOCLONAL ANTIBODY.

**maceration:** Separation of cells by enzymic digestion. *See* PROTOPLAST ISOLATION.

**Macerozyme:** A commercial preparation containing a mixture of enzymes that degrade plant cell walls; used in protoplast isolation. *See* PROTOPLAST ISOLATION.

**machine vision:** *See* IMAGE ANALYSIS.

**macroelements:** *See* MACRONUTRIENTS.

**macrolides:** A group of antibiotics that inhibit protein synthesis by bacteria at the 50S ribosome, e.g., erythromycin (Walsh, 2003). *See* ANTIBIOTICS.

**macronutrients:** Chemicals required in large concentrations for plant growth, including carbon, hydrogen, and oxygen obtained from carbon dioxide and water; nitrogen, potassium, phosphorous, sulfur, magnesium, calcium, and iron provided in culture media as calcium and potassium nitrates, potassium phosphate, iron phosphates or chelate, and magnesium sulfates (Marschner, 1994; Table 1). *See* MICRONUTRIENTS; MINERAL NUTRITION; PLANT TISSUE CULTURE MEDIA.

**macropropagation:** *See* VEGETATIVE PROPAGATION.

**Magenta container:** A commercial tissue culture vessel.

**magnesium:** Mg; essential nutrient, a component of chlorophyll and an enzyme cofactor (Marschner, 1994). *See* MACRONUTRIENTS.

**major gene resistance:** A gene conferring resistance that shows Mendelian inheritance (Agrios, 1997; Allard, 1999; Strange, 2003).

TABLE 1. Plant mineral nutrients.

| Mineral | Salt form in MS medium | Alernative salt form |
|---|---|---|
| Macronutrients | | |
| Nitrogen | $NH_4NO_3$, $KNO_3$ | $NH_4H_2PO_4$, $Ca(NO_3)$.$4H_2O$ $NH_4SO_4$, |
| Sulfur | $MgSO_4$.$7H_2O$ | $NH_4SO_4$, $K_2SO_4$, $Na_2SO_4$ |
| Phosphorus | $KH_2PO_4$ | $NaH_2PO_4$ |
| Magnesium | $MgSO_4$.$7H_2O$ | |
| Calcium | $CaCl_2$.$2H_2O$ | $Ca(N4H_2OO_3)$. |
| Potassium | $KNO_3$ | $KCl$, $K_2SO_4$ |
| Micronutrients | | |
| Iron | $FeSO_4$.$7H_2O$ | $Fe_2(SO_4)_3$ |
| Manganese | $MnSO_4$.$H_2O$ | |
| Copper | $CuSO_4$.$5H_2O$ | |
| Zinc | $ZnSO_4$.$7H_2O$ | |
| Nickel | Not present | |
| Molybdenum | $Na_2MoO_4$.$2H_2O$ | $NaMoO_3$ |
| Boron | $H_3Bo_3$ | |
| Chlorine | $CaCl_2$.$2H_2O$ | $KCl$ |
| Beneficial minerals | | |
| Sodium | $Na_2EDTA$ | |
| Silicon | Not present | |
| Cobalt | $CoCl_6$.$6H_2O$ | |
| Selenium | Not present | |
| Aluminum | Not present | |
| Other | $KI$ | |

*Source:* The categories are based on Marschner (1994). *Notes:* The materials used in tissue culture may be contaminated with trace elements; e.g., iodine is found as a contaminant of agar and the materials used may bind minerals (Cassells and Collins, 2000). Details of media mineral formulations are given in George (1993, 1996).

**maleic hydrazide:** A herbicide and plant growth inhibitor. It is used to inhibit sprouting in potatoes and onions, to induce dormancy in citrus, and as a plant growth regulator to control growth of a range of crops, including potatoes (George, 1993; Basra, 2000).

**malt extract:** A liquid extract from barley or other cereal grain allowed to sprout and then dried in a kiln. It consists of ~92 percent carbohydrate, 6 percent partially hydrolyzed protein, and 2 percent ash. It is a source of vitamins B1, B2, B6, B12, nicotinamide, pantothenic acid, folic acid, biotin, and ascorbic acid; of the macronutrients potassium, phosphorous, magnesium, calcium, and sodium; and of the micronutrients iron, copper, zinc, manganese, and chromium.

**maltose:** A sugar formed by the action of the enzyme diastase on starch (Bryant et al., 1999).

**manganese:** Mn; an essential micronutrient that is an enzyme cofactor (Marschner, 1994). *See* MICRONUTRIENTS.

**mannitol:** A common sugar alcohol derived from mannose or fructose (Bryant et al., 1999). It is the main soluble sugar in fungi (Watkinson and Gooday, 2001). Mannitol is used to plasmolyze plant cells in protoplast isolation (Warren, 1991) and as a component of slow growth and cryopreservation media (Cassells, 2002). *See* CRYOPRESERVATION; GENE BANKS; PROTOPLAST ISOLATION.

**mannose:** An aldohexose sugar stored in polymeric form as mannans in legumes (Bryant et al., 1999). Mannose is a component of some hemicelluloses (Brett and Waldron, 1989).

**marker gene:** A gene used to track an event such as transformation of a cell (Slater et al., 2003). *See* GENETIC ENGINEERING; β-GLUCURONIDASE GENE; GREEN FLUORESCENT PROTEIN.

**mass spectrometer:** Produces charged particles (ions) from the molecules that are to be analyzed using electric and magnetic fields to measure the mass of the charged particles (Grob and Barry, 2004). *See* GAS-LIQUID CHROMATOGRAPHY.

**maternal inheritance:** Inheritance of the organellar genomes (chloroplast and mitochondrion) from the female parent. Usually, pollen contains only the nuclear genes of the male parent (Grant, 1975; Allard, 1999). *See* CYBRIDS; PROTOPLAST FUSION.

**maturation:** The process of becoming mature (Howell, 1998). *See* PHASE CHANGE.

**mature phase:** The phase in plant development before which flowering cannot occur (Howell, 1998). It is associated with loss of rooting ability in woody species (Hartmann et al., 2001). *See* JUVENILE PHASE; PHASE CHANGE.

**MCPA:** A phenoxyacetic herbicide (Basra, 2000). *See* AUXINS.

**mDNA:** *See* MESSENGER DNA.

**media:** Nutritive substrates for the culture of microorganisms, cells, tissues, and organs in vitro. Media may be of known composition and/or may contain extracts of plant or animal origin, e.g., coconut milk (Lelliot and Stead, 1987; George, 1993, 1996; Schaad et al., 2001). *See* BACTERIOLOGICAL MEDIUM; PLANT TISSUE CULTURE MEDIA.

**media preparation:** The preparation of a stock solution of macro- and micronutrients, organic components, and plant growth regulators, involving the addition of water, adjustment of pH, addition of agar, and autoclaving. Alternatively, the components or medium may be purchased as a premixed powder. Usually, prepared formulations lack sucrose, agar, and plant growth regulators. Labile compounds should not be autoclaved but added after either autoclaving and cooling of the medium or sterile filtration (George, 1993; Rayns and Fowler, 1993; Beyl, 2005). *See* AUTOCLAVE; GOOD LABORATORY PRACTICE; STERILE FILTRATION.

**media sterilization:** Media are usually sterilized by autoclaving; heat-labile compounds are added by sterile filtration after cooling but before setting of the medium (George, 1993). *See* AUTOCLAVE; STERILE FILTRATION.

**media storage:** The storage of bulk or dispensed media. Bulk agar-solidified media may be melted using microwaves. Care should be taken to avoid desiccation of prepoured media. Stored media should be examined for contamination before use. Contaminants may indicate faulty equipment or procedures (Cassells and Doyle, 2004; Beyl, 2005). *See* GOOD LABORATORY PRACTICE; MICROWAVE OVEN.

**medicinal plants:** Plants that contain secondary metabolites with medicinal properties (Bonnett and Glasby, 1991). *See* CELL SUSPENSION CULTURE; SECONDARY METABOLITES.

**meiosis:** The process in which a diploid cell divides to give haploid gametes and during which the exchange of genetic material between the parental chromosomes occurs, followed by segregation of the recombinant chromosomes. Meiosis is divided into two stages, with the events of meiosis II resembling those of mitosis (Dyer, 1979; Russell, 2002; Mauseth, 2003). *See* MITOSIS.

**membrane filtration:** A process in which compounds are separated based on their size relative to the pore size of the membrane. Used to remove microorganisms from liquids, e.g., in the preparation of aseptic solutions of heat-labile compounds for addition to tissue culture media (George, 1993). *See* MEDIA PREPARATION; ULTRAFILTRATION.

**membrane raft:** A commercially available tissue support that floats above a liquid medium. The medium is conducted by wicks to the raft platform on which the tissue is placed. *See* TISSUE SUPPORTS.

**Menard's medium:** A meristem culture medium containing substrates to encourage bacterial growth, thereby facilitating the early detection of bacterial contamination (Menard et al., 1985).

**mercaptoethanol:** A reducing agent used to break disulfide bridges.

**mercuric chloride:** A broad-spectrum sterilant (George, 1993). Caution: mercury compounds are highly toxic in the environment and should be disposed of appropriately. *See* SURFACE STERILANTS.

**mericlinal chimera:** *See* CHIMERA.

**meristem:** A region containing mitotically active, or quiescent, undifferentiated cells that gives rise to the primary tissues of the plant (apical and lateral bud and root meristem) and lateral meristems (cambium), which give rise to secondary plant growth (Esau, 1977; Howell, 1998; Lyndon et al., 1998). *See* APICAL MERISTEM; MERISTEM CULTURE.

**meristem culture:** The culture of shoot-tip explants usually comprising the apical shoot tip and one or more pairs of leaf primordia (George, 1993). Pathogens affecting the plant may be restricted to the vascular system, and as this is differentiated below the apical tip, small apical explants may be pathogen free. This phenomenon is exploited in the use of apical explants to establish pathogen- and contaminant-free cultures (George, 1993; Faccioli and Marani, 1998; Kane, 2005). *See* MICROPROPAGATION; PATHOGEN ELIMINATION.

**meristem isolation:** Meristems are isolated using a scalpel to remove the bud scales progressively from the outside to expose the apical tip and first pairs of leaf primordia. The process is performed using a dissecting microscope (George, 1993; Kane, 2005).

**meristemoids:** Active loci of growth forming differentiated regions, e.g., root hairs, stomata, and procambial strands. Meristemoids are maintained at some distance from one another to form a fairly regular pattern (Cutter, 1978).

**meristem tip culture:** *See* MERISTEM CULTURE.

**messenger DNA:** mDNA; small pieces of DNA complexed with RNA and glycolipoprotein passing between cells and acting as a messenger (Gahan, 2003). *See* CYTOPLASMIC MALE STERILITY; ENDOSYMBIONT HYPOTHESIS.

**messenger RNA:** mRNA; RNA carrying information from gene to ribosome for protein synthesis. Protein synthesis occurs on mRNA on the ribosome (Soll et al., 2001; Alberts et al., 2002).

**metabolomics:** The use of advanced spectroscopic and spectrometric methods to analyze metabolites. May involve the use of artificial intelligence to analyze the data (van der Greef et al., 2004). *See* GENOMICS; PROTEOMICS.

**methylation:** *See* DNA METHYLATION.

**methylation-sensitive restriction enzymes:** Enzymes that cut DNA at specific methylated base sequences (Russell, 2002). *See* EPIGENETICS; RESTRICTION ENZYMES.

**methylcytosine:** *See* DNA METHYLATION.

**methyl jasmonate:** A volatile, naturally occurring growth regulator widely distributed in plants and active in pest stress signaling (Strange, 2003). *See* JASMONIC ACID.

**mevalonic acid:** A direct precursor of isopentenyl pyrophosphate, a precursor molecule for steroids, dolichols, tocophorols, phytol, chlorophyll, gibberellins, carotenoids, vitamins A and D, and gutta-percha (Towers and Stafford, 1990; Metzler, 2003).

**microarthropods:** *See* MITES.

**microbial contamination:** *See* GOOD LABORATORY PRACTICE.

**microbulb:** Bulb produced in vitro (Takayama et al., 1991).

**microelements:** *See* MICRONUTRIENTS.

**micrografting:** The grafting of tip explants to aseptically germinated seed or microplant stock in vitro. It is used to rejuvenate buds from mature phases of woody species, to establish virus-free shoots from infected plants, and to transmit viruses to indicator plants (George, 1993). *See* REJUVENATION.

**microhydroponics:** Aseptic photoautotrophic culture. This approach to tissue culture resembles hydroponic (soilless) culture in being based on the use of simple mineral media formulations, usually variants of Hoagland's solution, but unlike soilless culture, it is carried out under aseptic conditions as heterotrophic culture (Cassells, 2000b). *See* HETEROTROPHIC GROWTH; HOAGLAND'S SOLUTION; HYDROPONICS.

**microinjection:** A method used to introduce DNA or organelles into a cell via a thin glass needle or micropipette.

**micronutrients:** Chemical elements required in small quantities for plant growth that function as enzyme cofactors or essential components of pigments and enzymes. They include boron, cobalt, copper, molybdenum, manganese, and zinc. Some plants may require either chlorine or silicon or sodium or vanadium (See Marschner, 1994; Table 1). *See* MACRONUTRIENTS; MINERAL NUTRITION.

**microplant:** Syn. plantlet; plant produced in vitro. *See* MICROPROPAGATION.

**microplant establishment:** The physiology of tissues produced in vitro may be aberrant as a result of high humidity in the culture vessel and suppression of photosynthesis by high sucrose in the medium; consequently, microplants are frequently shaded and mist irrigated at weaning. Microplants are also susceptible to damping-off diseases and pest infestations but may react adversely to pesticides (Preece and Sutter, 1991; George, 1993, 1996; Debergh, et al., 2000). *See* IN VITRO WEANING; MICROPROPAGATION; RIBULOSE BISPHOSPHATE CARBOXYLASE.

**microplant quality:** Product quality is governed by the principle of "fitness for purpose." In the case of microplants, the propagules should be genetically true to type and of good physiological quality such that they have a 100 percent establishment rate and can be weaned without a growth lag. It is also important that they are developmentally normal (Preece and Sutter, 1991; Swartz, 1991; Lumsden et al., 1994; Grunewaldt-Stocker, 1997; Davies and Santamaria, 2000). *See* EPIGENETIC VARIATION; MICROPLANT ESTABLISHMENT; SOMACLONAL VARIATION; WEANING OF PLANTS.

**micropropagation:** The clonal propagation of plants in tissue culture. Generally the term refers to commercial large-scale propagation of plants in vitro (Debergh and Zimmermann, 1991; George, 1993; Pierik, 2002; Cassells, 2003; Figure 18; Exhibit 2). *See* ARTIFICIAL SEED; COMMERCIAL MICROPROPAGATION; SOMACLONAL VARIATION; STABLE CLONING.

**micropropagation laboratory:** A laboratory designed for micropropagation, i.e., designed to minimize the risks of laboratory contamination and containing the facilities and equipment required (Cassells, 1991; George, 1993, 1996; Beyl, 2005). *See* AUTOCLAVE; INSTRUMENT STERILIZATION; LAMINAR FLOW CABINET; MEDIA PREPARATION; WATER QUALITY.

**microsatellites:** Short tandem repeats (STRs) or simple sequence repeats (SSRs) consisting of tandomly repeated units, between one and ten base pairs in length, that are widely dispersed throughout the genome. They are used as markers in both plant breeding and identification (Ciofi et al., 1998). *See* GENETIC FINGERPRINTING.

**microspore:** The pollen grain in seed plants (Mauseth, 2003).

**microspore culture:** *See* POLLEN CULTURE.

**microsporogenesis:** Regeneration from pollen (Bhojwani and Soh, 2001).

**microtubers:** Small tubers produced in vitro. *See* SEED TUBERS.

**microwave oven:** A cooker used to liquefy small quantities of agar-solidified media.

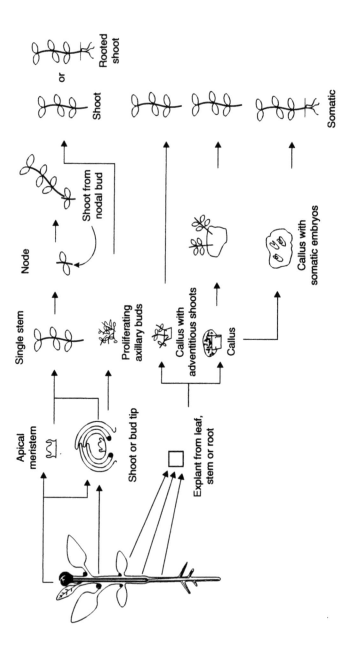

FIGURE 18. Strategies for cloning plants. There are two types of explants viz. explants containing apical meristems or explants lacking apical meristems such as leaf, stem, or root explants. The apical meristem or bud-containing explants can be grown into single stems and further propagated via nodal explants or induced to produce axillary bud clusters which are further subdivided. Adventitious shoots or somatic embryos can be induced directly in explants or indirectly via callus produced on the explant and subcultured. These events may be repeated until the desired number of progeny shoots or somatic seedlings is obtained. There is a risk of somaclonal variation when adventitious regeneration occurs; this is dependent on the genotype, age and tissue origin of the explant, media and environmental factors, and duration in culture (Mohan Jain et al., 1998). See CHIMERA, CLONING STRATEGY, MICROPLANT QUALITY, MICROPROPAGATION.

**mineral deficiency:** Usually defined by the expression of symptoms that are only expressed when the deficiency is acute (Marschner, 1994; Figure 19). *See* LEAF MINERAL ANALYSIS.

**mineral nutrition:** The essential macronutrients required by plants are nitrogen, phosphorous, sulfur, potassium, magnesium, and calcium; the essential micronutrients are iron, manganese, zinc, copper, boron, molybdenum, chlorine, and nickel. Nonessential but beneficial micronutrients are sodium, silicon, cobalt, iodine, and vanadium (Marschner, 1994). The requirements of plants for mineral nutrients vary, with deficiencies resulting in reduced growth and, at the acute stage, deficiency symptoms being expressed. Excess nutrient supply may result in toxicity. In the field, the crop nutrient requirements are based on soil mineral analysis and the application of fertilizers. In the growing season, leaf mineral analysis is used to determine fertilizer application. Mineral analysis of leaf tissue provides a good indication of the mineral requirements of the genotype and may be used to optimize the plant's requirements in vitro (Marschner, 1994). Most mineral nutrient formulations for hydroponics (including autotrophic culture) are based on Hoagland (1948) and those for heterotrophic plant tissue culture on Murashige and Skoog (1962). Factors affecting the availability of minerals in media are the pH and the presence of chelating agents to prevent precipitation of iron (Marschner, 1994), and uptake into the tissues may be dependent for some elements on a transpiration stream (Cassells and Roche, 1994; Cassells and Walsh, 1994).

**minimal growth storage:** A technique used for short-term storage of plant cultures involving growing at reduced temperature. The process may be associated with the inclusion of osmolytes such as mannitol in the medium (Cassells, 2002; Towill, 2005). *See* CRYOPRESERVATION.

**minitubers:** Tubers produced in a greenhouse from microplants in vivo. They are intermediate in size between microtubers and seed tubers. *See* MICROTUBERS; SEED TUBERS.

---

**EXHIBIT 2. Stages of micropropagation.**

Stage 0: The preparatory step in micropropagation involving the selection of stock plants and their maintenance under conditions that discourage pest and diseases. *See* DISEASE ELIMINATION; DISEASE INDEXING; PLANT HEALTH CERTIFICATION.

Stage 1: The establishment of aseptic cultures involving, normally, surface sterilization of the explant and, in the case of shoot tip explants, the dissection of the apical bud. *See* MERISTEM CULTURE.

Stage 2: The mass clonal propagation stage, involving the choice of an appropriate cloning strategy, indexing of the cultures. *See* CLONING; GOOD LABORATORY PRACTICE.

Stage 3: Preparation for return to the natural environment. This may involve shoot elongation, root induction, and manipulation of the culture vessel atmosphere to reduce humidity and improve calcium uptake by the shoots. *See* AUTOTROPHIC CULTURE; IN VITRO WEANING.

Stage 4: Establishment of the microplants, unrooted microshoots, or clumps of microshoots in a growth substrate in a greenhouse. Conventionally, the plants are shaded and mist-irrigated while they adapt their physiology to ambient conditions. Microplants may be inoculated with mycorrhizal fungi or plant-growth-promoting rhizobacteria at this stage. *See* BIOLOGICAL CONTROL; IN VITRO WEANING; MYCORRHIZAL FUNGI; PLANT-GROWTH-PROMOTING RHIZOBACTERIA.

Stage 5: Quality control checks on microplants after establishment to confirm health (good physiological status and freedom from disease) and developmental status. High-frequency mutation at unstable loci and epigenetic changes have been reported that are not detectable other than by visual examination of the phenotype. *See* EPIGENETICS; PLANT HEALTH CERTIFICATION.

*Reference:* George, E.F. (1993). *Plant Propagation by Tissue Culture, Part 1: The Technology.* Basingstoke: Exegetics.

---

**Mistifier:** Commercial apparatus designed for aseptic aeroponic culture of plant tissues in vitro. *See* HYDROPONICS.

**misting:** Irrigation of plants by application of vaporized water. Frequently used at microplant establishment (George, 1993). *See* MICROPLANT ESTABLISHMENT.

**mites:** Microscopic arthropods of the class Arachnida, usually having a rounded body with four pairs of jointed legs. They can cause serious infestations of plant growth rooms (Pype et al., 1997). *See* GOOD LABORATORY PRACTICE.

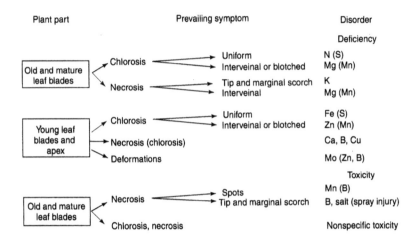

FIGURE 19. Symptoms of mineral deficiencies expressed *in vitro* (from Marschner, 1994). Reprinted by permission. Calcium deficiency is seen in some genotypes in sealed vessels (Cassells and Roche, 1994; Cassells and Walsh, 1994). *See* CALCIUM; HYPERHYDRICITY; TISSUE CULTURE VESSELS.

**mitochondria:** Double-membrane-bound organelles in the cytoplasm of eukaryotic cells. Mitochondria are the site of the tricarboxylic acid cycle and oxidative phosphorylation. They contain circular DNA and are capable of protein synthesis (Alberts et al., 2002; Mauseth, 2003). *See* ENDOSYMBIONT HYPOTHESIS; EUKARYOTES.

**mitochondrial DNA:** mtDNA; self-replicating, small circular DNA strands with no associated protein and coding for many proteins associated with the inner mitochondrial membrane (Alberts et al., 2002).

**mitosis:** The process of cell division to give two daughter cells with the same number of chromosomes and genetic composition as the parent cell, i.e., cell cloning (Russell, 2002). *See* ENDOMITOSIS; MUTATION; SOMACLONAL VARIATION.

**mitotic crossing over:** Genetic recombination in somatic cells (Russell, 2002).

**mixed cultures:** *See* COCULTURE.

**mixoploidy:** Cells or tissues with different chromosome numbers in the same individual (D'Amato, 1977). *See* CHIMERA; SOMACLONAL VARIATION.

**mixotrophic cultures:** Cultures whose energy for growth is partially derived from light and partially from carbon substrates in the medium. Although tissue cultures are commonly referred to as being heterotrophic, implying dependence on a carbon growth source, usually sucrose, in the medium, this overlooks the contribution of growth room PAR light to photosynthesis (Capellades et al., 1991). *See* AUTOTROPHIC CULTURE; HETEROTROPHIC GROWTH; RIBULOSE BISPHOSPHATE CARBOXYLASE.

**modular plug systems:** Planting trays separated into compartments. They are used in automated production (Hartmann et al., 2001).

**moisture vapor transmission rate:** MVTR; the rate at which water crosses a membrane; depends on the material in the membrane, the temperature, and the moisture gradient (Cassells and Roche, 1994; Cassells and Walsh, 1994). *See* IN VITRO ATMOSPHERE; TISSUE CULTURE VESSELS.

**molality:** The number of particles in a mass weight of fluid (e.g., millimoles per kilogram).

**molarity:** The number of particles of a particular substance in a volume of fluid (e.g., millimoles per liter).

**mole:** One mole is equal to $6.02 \times 10^{23}$ particles (Avogadro's number).

**molecular beacons:** Fluorescent probes for the identification of pathogens. It is possible to detect more that one pathogen in an extract by using different color fluorochromes simultaneously (Klerks et al., 2001). *See* PATHOGEN INDEXING.

**molecular hybridization probes:** RNA or DNA sequences that hybridize with DNA or RNA strands with complementary sequences. If the target sequence is known, such probes can be used to detect them (Karp et al., 1998). *See* BIOTIN LABELING; POLYMERASE CHAIN REACTION; RADIOLABELING.

**molecular mass:** The mass of a molecule is equal to the sum of the mass of all of the atoms present. This is equivalent to the molecular weight and is measured in daltons (Da) or kilodaltons (kDa), where 1 Da equals $1.661 \times 10^{-24}$ g.

**molecular weight:** Ratio of the weight of one molecule to one-sixteenth the weight of an oxygen atom.

**mollicutes:** Eubacteria with low guanine + cytosine content in their DNA. Mollicutes are gram-positive bacteria lacking the classical bacterial cell wall and were formerly called mycoplasmas and spiroplasmas. Pathogenic species are both phloem restricted and resistant to rifampicin. They are generally fastidious. *Spiroplasma* is cultivable; *Phytoplasma* is not cultivable (Bove and Garnier, 1997). *See* ANTIBIOTICS; FASTIDIOUS BACTERIA.

**molybdenum:** Mo; essential micronutrient involved in nitrate reduction and enzyme cofactor (Marschner, 1994). *See* MICRONUTRIENTS; MINERAL NUTRITION.

**monoclonal antibody:** MAb; antibody derived from the progeny of a single immune cell. A tumor cell that can replicate repeatedly is fused with mammalian cells that produce the antibody. The resultant hybridoma will produce antibodies continuously (Roitt et al., 2001; Coico et al., 2003). *See* ANTIBODY.

**monocot:** *See* MONOCOTYLEDONAE.

**Monocotyledonae:** Monocots, a subclass of the Angiospermae, containing all the flowering plants with one cotyledon. Monocots generally lack lateral meristems; they have scattered vascular bundles, narrow parallel-veined leaves, and a fibrous root system (Esau, 1977; Heywood, 1993; Mauseth, 2003). *See* DICOT.

**monosaccharides:** Polyhydroxyaldehydes or polyhydroxyketones with the general composition $(CH_2O)n$, e.g., glucose. They exhibit isomerism and have either L (laevo) or D (dextro) forms. Usually, only the D forms are used by cells (Metzler, 2001).

**morphogenesis:** The development of organs or parts of organs (Howell, 1998; Westhoff, 1998).

**morphogenic callus:** Undifferentiated cells that give rise to organs or parts of organs (George, 1993).

**morphogenic competence:** The state in which cells or tissues are able to respond to stimuli for regeneration of organs or parts of organs (George, 1993). *See* DETERMINATION.

**morphogenic potential:** *See* MORPHOGENIC COMPETENCE.

**morphology:** The shape and structure of an organism (Esau, 1977).

**mother plant:** *See* STOCK PLANT.

**mRNA:** *See* MESSENGER RNA.

**MS:** *See* MURASHIGE AND SKOOG MEDIUM.

**mtDNA:** *See* MITOCHONDRIAL DNA.

**multiplication:** Increasing the number. Multiplication usually implies cloning. *See* CLONING.

**multiplication rate:** The rate at which multiplication occurs. *See* COMMERCIAL MICROPROPAGATION.

**multistemming:** A phenomenon observed in vitro, associated with hyperhydricity, in which apical dominance is lost, resulting in development of lateral stems (Cassells and Roche, 1994). *See* HYPER-HYDRICITY.

**Multodisk:** A commercial system used in screening for antibiotic sensitivity. *See* ANTIBIOTICS.

**Murashige and Skoog medium:** MS medium; by far the most widely used formulation for plant tissue culture (~75 percent of the media cited in the literature) (Murashige and Skoog, 1962).

**mutagenesis:** The induction of mutation by chemical or physical methods (Mohan Jain et al., 1998; Van Harten, 1998). *See* GENETIC MANIPULATION; MUTATION BREEDING.

**mutant lines:** Clones selected from a mutation-breeding program (van Harten, 1998). *See* MUTATION BREEDING.

**mutation:** A heritable change in the amount or chemical structure of the DNA of the organism that results in the nonexpression of proteins or alteration in proteins expressed by the organism (Russell, 2002). *See* CHEMICAL MUTAGENS; PHYSICAL MUTAGEN.

**mutation breeding:** Historically, the use of physical and chemical mutagens to increase variability in a plant breeding program; now expanded to include the use of physical and chemical mutagens in vitro and spontaneous mutation in vitro ("somaclonal variation") (Cassells et al., 1993; Mohan Jain et al., 1998; van Harten, 1998; Cassells, 2002). *See* SOMACLONAL VARIATION.

**MVTR:** *See* MOISTURE VAPOR TRANSMISSION RATE.

**mycoplasma:** *See* MOLLICUTES.

**mycorrhizal fungi:** Symbiotic root-inhabiting fungi that assist the plant in taking up phosphorous and water (Kapulnik and Douds, 2000; Poincare et al., 2002; Saxena and Johri, 2002; Duffy and Cassells, 2003). Mycorrhizal inoculants are used to protect microplants against weaning stresses resulting from damping-off pathogens and water stress (Vestberg et al., 2002). *See* ARBUSCULAR MYCORRHIZAL FUNGI; ECTOMYCORRHIZAL FUNGI.

**mycorrhizas:** Associations between fungi and plant roots (Duffy and Cassells, 2003). *See* ARBUSCULAR MYCORRHIZAL FUNGI; ECTOMYCORRHIZAL FUNGI.

**Mycostatin:** A mycocide, the commercial formulation of nystatin. *See* ANTIMYCOTICS; FUNGICIDES; NYSTATIN.

**myoinositol:** *See* INOSITOL.

 **n:** Haploid set of chromosomes (Grant, 1975).

**2n:** Diploid set of chromosomes (Grant, 1975; Dyer, 1979).

**N:** *See* NITROGEN.

**NAA:** *See* NAPHTHALENE ACETIC ACID.

**NAD:** Nicotinamide adenine diucleotide.

**NADP:** Nicotinamide adenine diucleotide phosphate.

**nanometer:** nm; one-thousandth of a micrometer ($10^{-9}$ m).

**1-naphthaleneacetamide:** Synthetic auxin used occasionally in tissue culture media (George, 1993). *See* AUXINS.

**naphthalene acetic acid:** 1-Naphthalene acetic acid, NAA; a synthetic auxin (Basra, 2000) capable of promoting rooting and regarded as being more potent than indoleacetic acid (IAA) in tissue culture (George, 1993). *See* AUXINS.

**2-naphthoxyacetic acid:** Synthetic auxin capable of promoting rooting in cuttings (George, 1993). *See* AUXINS.

**NAPPO:** *See* NORTH AMERICAN PLANT PROTECTION ORGANIZATION.

**natural selection:** Selection of organisms with the genotypes most suited to a particular environment (Russell, 2002).

**necrosis:** Death of one or a group of plant cells in the living plant, often forming a circular patch (Strange, 2003). *See* LOCAL LESION.

**negative staining:** Use of dyes to enhance the background, thus showing up the contours of the specimen. Negative staining is

doi:10.1300/5648_14

especially used in electron microscopy, with contrast being induced with electron-dense phosphotungstic acid and sodium molybdate (Dijksta and de Jeger, 1998; Hari and Das, 1998).

**neomycin phosphotransferase:** NPTII; an enzyme that inactivates a number of aminoglycoside antibiotics such as kanamycin, neomycin, geneticin (or G418), and paramomycin by phosphorylation, which blocks the interaction between the antibiotic and ribosomes. These antibiotics are used in a wide range of plant species; however, kanamycin has proved to be ineffective in selecting for transformed legumes and monocots (Slater et al., 2003). *See* SELECTABLE MARKER.

**neotype:** When all type material for a species has been lost, a plant specimen is chosen that conforms as closely as possible to the original description.

**net assimilation rate:** Relationship of plant productivity to plant size. The rate is calculated as the rate of dry weight increase divided by leaf size.

**Ni:** *See* NICKEL.

**niacin:** *See* NICOTINIC ACID.

**niacinamide:** *See* NICOTINAMIDE.

**nickel:** Ni; metal sometimes included in tissue culture media (George, 1993; Marschner, 1994). *See* MICRONUTRIENTS; MINERAL NUTRITION.

**nicotinamide:** Syn. niacinamide; amide of niacin used occasionally in plant tissue culture medium (George, 1993). *See* PLANT TISSUE CULTURE MEDIA.

**nicotinic acid:** Syns. niacin, Vitamin B3; a member of the B vitamins, a component of the cofactors nicotinamide adenine dinucleotide (NAD) and nicotinamide adenine dinucleotide phosphate (NADP). *See* PLANT TISSUE CULTURE MEDIA.

**nipple flask:** System for the agitation and aeration of suspension cultures in which the medium can be continuously exchanged during the culture period (George, 1993).

**nitrate:** $NO^{3-}$; salt or ester of nitric acid (Marschner, 1994) added to all plant culture media as a source of nitrogen. *See* MACRONUTRIENTS; PLANT TISSUE CULTURE MEDIA.

**nitrogen:** N; gas making up ~78 percent of air. Nitrogen is formed in nitrogen fixation or synthetically. Nitrogen is an essential component of tissue cultures, provided as inorganic salt (nitrate or ammonium), urea, or organic nitrogen (amino acids, amides, peptides) (George, 1993). It is found in amino acids, proteins, nucleic acids, chlorophyll, and plant growth regulators. It is the element required in the greatest quantity for plant growth (Metzler, 2003).

**nitrogen fixation:** Atmospheric nitrogen is converted into ammonia, nitrites, and nitrates by blue-green algae and some bacteria (Postgate, 1998). *See* NITROGEN.

**Nitsch and Nitsch medium:** A medium developed for pollen culture (Nitsch and Nitsch, 1969).

**nodal culture:** Culture of lateral bud together with a portion of stem. This is the preferred explant in micropropagation to avoid somaclonal variation (George, 1993). *See* MICROPROPAGATION.

**node:** Site of either bud or leaf attachment to the stem (Esau, 1977).

**nodules:** *See* ROOT NODULES.

**noncompetent:** Opposite of competent. The term is used to describe cells unable to undergo morphogenesis (George, 1993). *See* COMPETENCE.

**nondisjunction:** Failure of paired chromosomes at meiosis I, or of sister chromatids in meiosis II, to separate and move to opposite poles, resulting in one daughter cell getting both and the other neither. of the chromosomes involved (Dyer, 1979; Russell, 2002).

**North American Plant Protection Organization:** NAPPO; one of the international organizations that implements regional plant health certification on behalf of the FAO. *See* EUROPEAN AND MEDITERRANEAN PLANT PROTECTION ORGANIZATION; FOOD AND AGRICULTURE ORGANIZATION OF THE UNITED NATIONS.

**NPA:** 1-Naphthylphthalamic acid.

**NPTII:** *See* KANAMYCIN RESISTANCE; NEOMYCIN PHOSPHOTRANSFERASE.

**nucellar embryony:** Production of embryos from individual cells of the nucellus.

**nucellus:** Mass of parenchymatous tissue within the ovule containing the embryo sac and surrounded by the integuments (Esau, 1977).

**nucleic acid:** *See* DEOXYRIBONUCLEIC ACID; RIBOSE NUCLEIC ACID.

**nucleo-cytoplasmic ratio:** Ratio of nuclear to cytoplasmic volume.

**nucleolar organizer:** Portion of the DNA coding for ribosomal RNA genes (Alberts et al., 2002). *See* NUCLEOLUS.

**nucleolus:** Contains rRNA genes, precursor rRNAs, mature rRNAs, small nucleolar ribonucleoprotein particles (snoRNPs), ribosomal subunits, partly assembled ribosomes, and enzymes involved in RNA polymerization. The nucleolus, which is not delimited by a membrane, is the site of the biosynthesis of ribosomal and transfer RNA and ribosome production (Alberts et al., 2002; Thompson et al., 2003). *See* NUCLEOLAR ORGANIZER.

**nucleoside:** Pentose sugar linked to either a pyrimidine or a purine (Metzler, 2003).

**nucleosome:** Two molecules each of histones H2A, H2B, H3, and H4 in a structure 11 nm in diameter around which is coiled DNA as a first stage in the packaging of a chromosome in the interphase nucleus (Alberts et al., 2002; Russell, 2002). *See* HISTONE; HISTONE ACETYLATION.

**nucleotide:** A nucleoside esterified to phosphate at the C5 of the pentose sugar (Metzler, 2001).

**nucleus:** Eukaryotic cell structure containing the chromosomes. The nucleus is surrounded by a double membrane containing pores for exchange of material between nucleus and cytoplasm, the outer membrane being continuous with the endoplasmic reticulum (Alberts et al., 2002; Russell, 2002). *See* EUKARYOTES; PROKARYOTE.

**nullisomic:** Aneuploid chromosome complement with both members of a chromosome pair missing from the diploid set (Dyer, 1979; Allard, 1999).

**nurse culture:** Placing a single cell or cells on filter paper or membrane in contact with a callus (nurse material) from which the cell derives nutrients and other factors for development (Warren, 1991). *See* PROTOPLAST ISOLATION.

**nutrient:** Components of culture media required to sustain the growth and development of plants and of plant tissues in vitro. *See* MINERAL NUTRITION; PLANT TISSUE CULTURE MEDIA.

**nutrient film technique:** *See* HYDROPONICS.

**nutrient gradient:** Diffusion gradient of nutrients and gases within an explant placed in contact with a culture medium.

**nystatin:** A mycocide used in protoplast isolation. *See* FUNGICIDES.

**off-types:** In vegetative propagation, plants that are not clones of the parent (Hartmann et al., 2001). *See* CHIMERA; CLONE.

**ontogeny:** The origin and the development of an organism from the fertilized egg to its mature form (Howell, 1998; Graham et al., 2003).

**organ:** A part of an organism adapted for a specific function, e.g., the leaf for photosynthesis (Esau, 1997; Mauseth, 1988).

**organelle:** A membrane-bound structure in eukaryotes in which specific functions are performed (Taiz and Zeigler, 2002; Graham et al., 2003; Mauseth, 2003).

**organic supplements:** Organic compounds added to a basal medium, e.g., coconut milk (George, 1993). *See* PLANT TISSUE CULTURE MEDIA.

**organized growth:** Growth showing differentiation into tissues and organs. *See* CALLUS TISSUE.

**organogenesis:** The formation of organs. *See* ADVENTITIOUS REGENERATION.

**ornamentals:** Plants grown for decorative purposes (Hartmann et al., 2001).

**ortet:** The stock or mother plant.

**orthotropic:** A growth response in which the response is toward or away from the stimulus (Taiz and Zeigler, 2002). *See* PLAGIOTROPIC.

**oryzalin:** A microtubule-disrupting drug used in tissue to produce allopolyploids. *See* ALLOPOLYPLOIDS; COLCHICINE.

doi:10.1300/5648_15

**osmolality:** The total number of osmotically active particles in a solution. Osmolality is equal to the sum of the molalities of all the solutes present per kilogram of solvent. For most biological systems the molarity and the molality of a solution are almost exactly equal because the density of water is 1 kg·L$^{-1}$. *See* OSMOLARITY.

**osmolarity:** Proportional to the number of particles per liter of solution. *See* OSMOLALITY.

**osmolytes:** *See* OSMOTICUM.

**osmoregulation:** The active regulation of the osmotic pressure of the cytoplasm to maintain the homeostasis of the cell's water content (Ridge, 2002).

**osmotic potential:** Syn. water potential; the tendency of water to enter or to leave a cell or tissue. Water moves from a region of higher osmotic potential to one of lower osmotic potential. The osmotic potential of pure water is zero. All solutions have lower osmotic potential than pure water and therefore have negative values; the stronger a solution, the more negative its osmotic potential. When a cell is placed in a solution, water will always move from the solution at low osmotic potential to the cell, causing the cytoplasm to expand (Ridge, 2002). *See* HYPERTONIC; HYPOTONIC; PROTOPLAST ISOLATION.

**osmotic pressure:** The pressure necessary to reverse osmosis and return to the initial condition. The osmotic pressure at a given temperature depends upon the molar concentration.

**osmoticum, pl. osmotica:** Compounds used to adjust osmolarity (Warren, 1991).

**outbreeding:** The production of offspring by the crossing of distantly related gametes (Grant, 1975; Allard, 1999). *See* INBREEDING; INCOMPATIBILITY.

**ovule culture:** The culture of a plant female gamete; cocultivation with pollen is used to achieve wide crosses. Callus from ovules can be embryogenic (Beasley, 1977). *See* IN VITRO POLLINATION.

**oxidation:** The total number of electrons removed from an element (a positive oxidation state) or added to an element (a negative oxidation state) to get to its present state. Oxidation involves an increase in oxidation state; reduction involves a decrease in oxidation state.

**oxidative stress:** Cell damage caused by oxidizing agents (Figure 1), which include products of photosynthesis and respiration and of the activity of such enzymes as lipoxygenase. A wide range of free radicals derived from oxygen (ROS; reactive oxygen species) cause damage. Most frequently implicated are superoxide and hydroxyl ions, hydrogen peroxide, and singlet oxygen. Other chemical species involved include hypochlorous acid, peroxynitrite, and transition metal ions. Free radicals react with host lipids, e.g., in cell membranes, initiating peroxidation of lipids and causing damage to both proteins and DNA to induce mutation; these effects can be cytotoxic. Oxidative stress can be detected or measured by assaying markers for lipid peroxidation or by monitoring the depletion of antioxidants (Inze and van Montague, 2001). Oxidative stress has been implicated in somaclonal variation and hyperhydricity (Cassells and Curry, 2001; Gaspar et al., 2002; Joyce et al., 2003). *See* ABIOTIC STRESS; BIOTIC STRESS; FREE RADICALS; MUTATION.

**oxytetracycline:** Syn. Terramycin; a yellow crystalline broad-spectrum antibiotic produced by *Streptomyces rimosus* (Walsh, 2003) used for the control of plant bacterial and mycoplasma-related diseases such as fire blight, bacterial leaf scorch, elm/ash yellows, elm phloem necrosis, and lethal yellows of palm. It has been used to treat stock plants to eliminate *Xanthomonas* (Cassells et al., 1988). *See* ANTIBIOTICS.

**P**

**packed cell volume:** A measure of growth in cell cultures based on slow-speed centrifugation ($4000g \times 5$ min) of the cells in a calibrated centrifuge tube (Fowler and Rayns, 1993). *See* CELL SUSPENSION CULTURE.

**paclobutrazol:** A growth inhibitor, trade name Bonzi. Paclobutrazol is an inhibitor of gibberellin biosynthesis and inhibits P450 monooxygenases (Basra, 2000). It is used as an inhibitor of gibberellins in plant tissue cultures (George, 1993). *See* GIBBERELLIN INHIBITORS; PLANT GROWTH REGULATOR.

**palisade:** Layer of parenchymal cells beneath the adaxial epidermis of a mesophyte leaf or beneath both epidermises of xerophyte leaves. It contains chloroplasts (Esau, 1977).

**panicle:** Form of inflorescence in which the main axis bears racemes (several alternate or spirally arranged branches), each one bearing one or more flowers (Heywood, 1993).

**pantothenic acid:** Vitamin B5; widespread precursor of coenzyme A in plants (Metzler, 2001). It is added to culture medium as the calcium salt (George, 1993). *See* PLANT TISSUE CULTURE MEDIA.

**papain:** A cysteine or thiol protease present in papaya fruit (Metzler, 2001).

**paper bridges:** *See* FILTER PAPER BRIDGES.

**paper raft method:** Single cells are placed on a square of filter paper that is, in turn, placed on an actively growing callus (nurse culture). The cells receive growth factors and nutrients through the filter paper from the callus (Warren, 1991). *See* NURSE CULTURE; TISSUE SUPPORTS.

**PAR:** *See* PHOTOSYNTHETICALLY ACTIVE RADIATION.

doi:10.1300/5648_16

**para-aminobenzoic acid:** PBA; vitamin Bx, which acts as an anti-metabolite to sulfanilamide, blocking its action. It is occasionally added to tissue culture media (George, 1993).

**para-chlorophenoxyacetic acid:** PCPA; auxin-like synthetic growth regulator sometimes used in plant tissue culture media (George, 1993; Basra, 2000). *See* AUXINS.

**paraffin wax:** Relatively low-melting-point, translucent, white solid hydrocarbon used frequently as an embedding medium for microtome sectioning of plant material for light microscopy (Gahan, 1984).

**parafilm:** Stretchable waxed film used for covering glassware and sealing petri/culture dishes.

**Paraplast:** Brand of paraffin-based embedding medium for histology.

**parenchyma:** Often vacuolate cells surrounded by only a primary cell wall retaining the ability for cell division although mature. Parenchyma is considered to be the basic cell type from which all other cell types evolved and which may, therefore, be considered to be undifferentiated (Esau, 1977; Trigiano and Gray, 2005).

**PAR meter:** An instrument to measure photosynthetically active radiation. PAR light is measured in the range 400-700 nm. *See* ARTIFICIAL LIGHT; GROWTH ROOM LIGHTING; PHOTOSYNTHESIS.

**parthenocarpy:** Fruit formation without seed setting (Grant, 1975).

**parthenogenesis:** Development of a plant from an ovule that, owing to faulty meiosis, is diploid and so gives rise to diploid offspring (Grant, 1975).

**partial immersion:** Not submerged. *See* TEMPORARY IMMERSION.

**particle bombardment:** Syn. biolistics; a technique for introducing DNA into cells based on the shotgun principle. Developed for genetic transformation of monocots, which were thought to be recalcitrant to

*Agrobacterium* transformation (Christou, 1995, 1996; Gray, Compton et al., 2005). *See AGROBACTERIUM*-MEDIATED GENE TRANSFER; GENETIC ENGINEERING; PARTICLE GUN.

**particle gun:** A device for shooting DNA-labeled particles into cells. The particles are discharged under helium pressure (Gray, Compton et al., 2005; Figure 20). *See* BIOLISTICS; GENETIC ENGINEERING.

**partition chromatography:** Components of a mixture can be separated on a surface based on their partition coefficients using specific solvents (Fried and Sherma, 1999). *See* CHROMATOGRAPHY.

**passage:** Subculturing. The term is used in virology and microbiology.

**patent:** Legal protection for an invention can now be obtained for new plant varieties (Berman, 2002). *See* PLANT BREEDERS' RIGHTS; PLANT PATENT.

**pathogen:** Agent causing a disease (Agrios, 1997; Strange, 2003). *See* VITROPATHOGEN.

**pathogen elimination:** Dependent on the type of pathogen. In principle, meristem culture, depending on the size of the apical tip explant, can be used to eliminate most plant pathogens (Cassells, 2000a). The exceptions are viruses and viroids that penetrate the apex beyond the areas of vascular differentiation (Hull, 2002). Bacterial plant pathogens are also generally eliminated by meristem culture. Antiviral chemicals, antibiotics, and antimycotics/fungicides are also used in vitro to eliminate viruses, bacteria, and fungi, respectively. Thermotherapy has also been used in vitro (Lopez-Delgado et al., 2004). Some researchers, especially those working with woody species, prefer to apply thermotherapy to the stock plant and to initiate cultures from in vivo tested material (Mink et al., 1998; Cassells and O'Herlihy, 2003). *See* BACTERIAL ELIMINATION; PATHOGEN FREE; PATHOGEN INDEXING; PLANT HEALTH CERTIFICATION; VIROID ELIMINATION; VIRUS ELIMINATION.

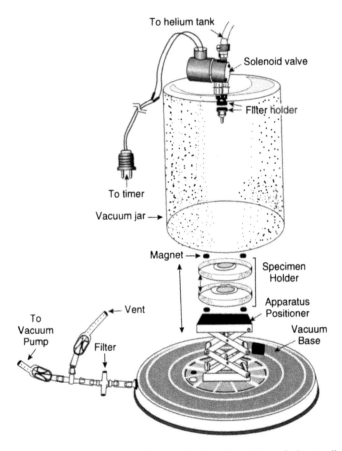

FIGURE 20. Illustration of a particle gun for transformation of plant cells (from Gray et al., 2005). Reprinted by permission of Springer Science and Business Media. *See AGROBACTERIUM* TRANSFORMATION; PROJECTILE BOMBARDMENT.

**pathogen free:** Term loosely used for "free of disease." This depends on the diseases for which the plant was tested and, in the case of official health certification, on the diagnostic methods used and the tissue to which they are applied (Cassells, 1997). *See* PLANT HEALTH CERTIFICATION.

**pathogen indexing:** Testing of plants for the presence of plant pathogens (Figures 21 and 22; Table 2). *See* PLANT HEALTH CERTIFICATION.

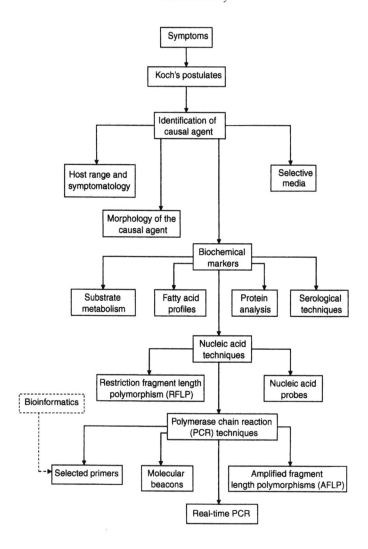

FIGURE 21. Methods for the identification of plant pathogens, from classical to molecular methods (based on Strange, 2003). *See* Table 2, *also* PATHOGEN INDEXING; PLANT HEALTH CERTIFICATION.

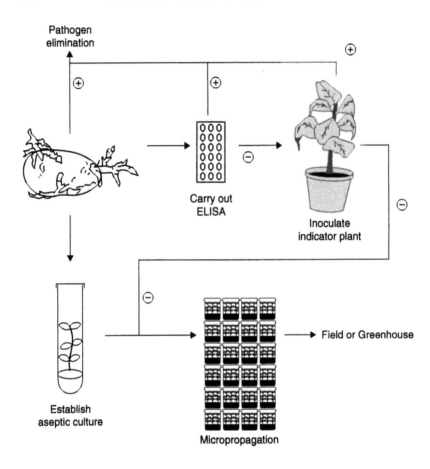

FIGURE 22. Pathogen indexing of potato stock plants and multiplication of pathogen-free cultures. Aseptic cultures are established from individual tubers via meristem culture. In parallel, the tuber is tested for pathogens by ELISA and by inoculation of indicator plants. In the case of ELISA negative results it is advisable to confirm pathogen absence by inoculation of an indicator plant. The meristem cultures of pathogen-negative tubers are then multiplied and the progeny microplants are planted in greenhouses or in the field for further multiplication prior to certification (Cassells and O'Herlihy, 2003). Tubers testing positive for pathogens in ELISA or indicator plants may be sent for virus elimination. *See* ELISA; INDICATOR PLANT; PLANT HEALTH CERTIFICATION.

TABLE 2. Application of indexing methods to the identification of plant pathogens.

| Pathogen type | Test methods | | | | | | | |
|---|---|---|---|---|---|---|---|---|
| | Symptoms | Indicator plants | Serological diagnostics | Microscopy | Biochemical test kits | Selective culture media | Genomic diagnostics | Proteomic diagnostics |
| Virus | Widely used | Widely used to confirm pathogenicity and identify strains | ELISA is standard method | Limited use of electron microscopy | Not applicable | Not applicable | Detection of variable coat genes/conserved sequences | Detection of viral coat proteins |
| Viroid | Widely used | Widely used to confirm pathogenicity and identify strains | Not applicable | Not used | Not applicable | Not applicable | Standard method | Not applicable |
| Bacterium[a] | Widely used | Widely used to confirm pathogenicity | Not commonly used | Widely used for gram and spore staining, etc. | Not commonly used for plant pathogens | Historically used in plant pathology | Detection of rDNA | Detection of bacterial biomarkers |
| Fungus | Widely used | Widely used to confirm pathogenicity | Not commonly used | Widely used for morphological identification | Not applicable | Widely used | Detection of rDNA | Detection of fungal biomarkers |

*Note:* Genomics methods are increasingly being used, and proteomic methods are also being applied.
[a] Fatty acid analysis is used for the identification of bacteria plant pathogens.

**pathogenesis-related proteins:** Proteins induced in plants following recognition by the plant that it is being attacked by pathogens or pests. These proteins include cellulases, glucanases, chitinases, proteinases, and some as yet uncharacterized proteins (Van Loon, 2000).

**PBA:** N-(phenylmethyl)-9-(tetrahydro-2*H*-pyran-2-yl)-9*H*-purin-6-amine; highly active synthetic cytokinin often used in plant culture media to induce either axillary bud or callus proliferation (George, 1993). *See* CYTOKININS.

**PCR:** *See* POLYMERASE CHAIN REACTION.

**peat:** Partially decomposed plant material. Peat accumulates in areas with poor drainage. Moss peat is acidic, nutrient poor, and rich in pathogen-antagonistic microorganisms; fen peat is less acidic and more nutrient rich. Both are used in horticultural potting composts (Hartmann et al., 2001).

**pectin:** High-molecular-weight methylated polymers of galacturonic acid present as magnesium and calcium salts in the middle lamella between primary cell walls and in some fruits (Bryant et al., 1999; Carpita et al., 2001). *See* PROTOPLAST ISOLATION.

**pectinase:** An enzyme. Pectinase is usually part of a cocktail of enzymes used to degrade pectin in protoplast isolation (Warren, 1991). *See* PROTOPLAST ISOLATION.

**PEG:** *See* POLYETHYLENE GLYCOL.

**penicillin:** A $\beta$-lactam antibiotic from *Penicillium notatum* and related species that inhibits bacterial wall synthesis (Walsh, 2003). *See* ANTIBIOTICS.

**pentose phosphate pathway:** Hexose monophosphate shunt in which glucose is oxidized to form 3-, 4-, 5-, and 7-carbon sugars as intermediates for a range of anabolic processes. Most steps do not require energy and the $NADPH_2$ produced is important in lipid synthesis (Bryant et al., 1999; Metzler, 2003).

**pentose sugar:** A monosaccharide containing five carbon atoms, e.g., ribose, deoxyribose, ribulose, xylose, xylulose (Bryant et al., 1999). *See* HEXOSE SUGAR.

**peptide:** From 2 to 20 amino acids covalently linked (Buchanan et al., 2002). *See* POLYPEPTIDE.

**peptone:** Polypeptides derived from protein by proteolytic degradation. Used in bacteriological media (Schaad et al., 2001). *See* BACTERIOLOGICAL MEDIUM.

**periclinal chimera:** *See* CHIMERA.

**periclinal division:** *See* APICAL MERISTEM; CHIMERA.

**pericycle:** Outermost layer of the stele lying within the endodermis. From the pericycle are initiated lateral roots arising from the main root (Esau, 1977; Davis and Hassig, 1994).

**perlite:** Unique volcanic mineral that expands to between 4 and 20 times its original volume when quickly heated to a temperature of ~850-950°C. When expanded, each particle of perlite is sterile, has a neutral pH, and contains many tiny closed cells or bubbles. The surface of each particle is covered with tiny cavities, providing an extremely large surface area to hold moisture and nutrients and make them available to plant roots. In addition, because of the physical shape of each particle, air passages are formed that provide optimum aeration and drainage. Perlite is used as a component of growing mixtures with, e.g., peat or bark (Hartmann et al., 2001). It offers a superior growing medium for hydroponic cultures (Sholto-Douglas, 1986). *See* HYDROPONICS.

**peroxisome:** Syn. glyoxysome; a single-membrane-bound cytoplasmic organelle within the cytoplasm in which up to 50 enzymes may be found, including peroxidase, catalase, and urate oxidase. The glyoxylate cycle, $\beta$-oxidation of very long fatty acids, photorespiration pathways, and synthetic pathways for cholesterol and plasmalogens are located in the peroxisome (Metzler, 2003).

**pesticides:** Compounds that kill pests. The term is used generally for herbicides, insecticides, and fungicides (Matthews, 1999; Basra, 2000; Bohmont, 2002).

**petal:** A modified leaf forming an individual component of the corolla (Esau, 1977). Occasionally used as an explant for adventitious regeneration in recalcitrant plants (George, 1996). *See* ADVENTITIOUS REGENERATION.

**petiole:** The leaf stalk linking the leaf blade to the stem (Esau, 1977). It is widely used as an explant for adventitious regeneration and callogenesis (George, 1996; Trigiano and Gray, 2005). *See* ADVENTITIOUS REGENERATION.

**petri dish:** A shallow glass or plastic covered dish usually of 5, 9, or 14 cm diameter.

**PGPR:** *See* PLANT-GROWTH-PROMOTING RHIZOBACTERIA.

**pH:** The negative logarithm of the hydrogen ion concentration, yielding a scale of 0-14 with a 10-fold difference between each unit of change. pH gives a measure of the acidity or alkalinity of a solution; pH 7.0 is neutral, with lower values indicating increasing acidity and higher values increasing alkalinity. The solubility of plant macro- and micronutrients is pH dependent (Marschner, 1994). *See* PLANT TISSUE CULTURE MEDIA.

**phase change:** One of several discrete stages in plant development (Poethig, 1990; Lawson and Poethig, 1995). Although heterochrony may be seen in the development of individual organs such as leaves, it is the juvenile-adult reproductive phase changes of the plant that are most important for the ability to vegetatively propagate (Hartmann et al., 2001) and for tissue culture (George, 1993). Constant plant morphology during the vegetative phase makes it difficult to recognize juvenile to adult phase change compared with the change from the vegetative to reproductive phase. It has long been known that many plant species show heteroblasty during vegetative development. In many cases (e.g., ivy), heteroblasty is expressed in heterophylly and is interpreted as a juvenile-adult phase change. In herbaceous plants, morphological, anatomical, and physiological traits change temporally, reflecting the juvenile-adult phase change. In maize, several traits such as epicuticular wax, epidermal features, and aerial roots characterize the juvenile phase (Poethig, 1990). It is hypothesized that most higher plants, if not all, have a genetic program for juvenile-adult phase change (Howell, 1998; Poethig, 2003). *See* HETERO-BLASTY; HETEROPHYLLY.

**phase contrast microscope:** A form of light microscope that converts (invisible) differences of phase in the transmitted light to (visible) changes of contrast. Thus a living, unstained specimen will have changes of light as a result of retardation of light wavelengths passing through more dense cell structures compared with the direct light waves; the two sets of waves interfere to give an image of varying degrees of gray, thereby allowing visualization of cell components (Gahan, 1984; Murphy, 2001).

**phellem:** The compact protective outer layer in plants exhibiting secondary growth and replacing the epidermis; it is composed of isodiametric cells with suberized walls (Esau, 1977). *See* SUBERIN; WOUND RESPONSE.

**phelloderm:** The secondary cortex, consisting of parenchyma cells derived from the phellogen (Esau, 1977).

**phellogen:** A cork cambium, a layer of secondary meristematic cells external to the vascular cambium and giving rise to the phelloderm and phellem (Esau, 1977).

**phenocopy:** Phenotype of one genotype resembling the phenotype of another genotype.

**phenolics:** *See* PHENOLS.

**phenols:** Organic compounds with hydroxyl groups attached to a benzene ring to form esters, ethers, and salts (Hemingway and Laks, 1992). Phenols are plant secondary products that protect against abiotic and biotic stress (Strange, 2003). *See* ABIOTIC STRESS; BIOTIC STRESS.

**phenotype:** The genetic expression of external characteristics setting the expression norm for environmental influences during an organism's development (Allard, 1999).

**phenotypic variation:** Variation in the phenotype. *See* PHENOTYPE.

**phloem:** Vascular tissue comprising sieve tubes, sclerenchyma, and parenchyma cells including companion cells, the function of which is the translocation of sugars and other nutrients. It occurs with the xylem, to which it is usually external (Wooding and Head, 1978). Viruses and bacteria move in the phloem (Hull, 2002). *See* PHLOEM-RESTRICTED MICROORGANISMS; PLANT VIRUSES.

**phloem-restricted microorganisms:** Bacteria and larger viruses that are too big to pass through the plasmodesmata into the cortical cells (Bove and Garnier, 1997; Hull, 2002). *See* FASTIDIOUS BACTERIA.

**phloroglucinol:** A glucoside of phloretic acid common in fruit trees and used in histology, after acidification, to temporarily react with lignin to give a red color. Promotes rooting in some woody species (George, 1993).

**pH meter:** A pH-measuring device.

**phosphate:** The form in which phosphorus is provided as a plant macronutrient (George, 1993; Marschner, 1994). *See* MINERAL NUTRITION; PLANT TISSUE CULTURE MEDIA.

**phospholipid:** A lipid comprised of glycerol, fatty acid, phosphate and an alcohol; may contain a nitrogen base and the common membrane lipids phosphatidylethanolamine, phosphatidylcholine, phosphatidylglycerol, and phosphatidylinositol (Metzler, 2003).

**phosphorus:** P; an essential macronutrient for plants. Phosphorus is a component of nucleic acids, lipoproteins, and energy compounds, e.g., adenosine triphosphate (ATP). It is relatively insoluble in soils (Marschner, 1994). *See* MYCORRHIZAL FUNGI.

**photoautotrophic culture:** *See* AUTOTROPHIC CULTURE.

**photoautotrophs:** Green plants and photosynthetic bacteria employing sunlight as their principle source of energy. *See* AUTOTROPHIC CULTURE; HETEROTROPHIC GROWTH; MIXOTROPHIC CULTURES.

**photobleaching:** The loss of pigments on exposure to excessive light (Ridge, 2002).

**photoheterotroph:** A photosynthetic organism requiring organic compounds as a source of hydrogen.

**photoinhibition:** The effect of excessive light by which the efficiency and rate of photosynthesis are reduced (Ragavendra, 1998; Rao, 1999). *See* PHOTOSYNTHESIS.

**photometer:** An instrument for measuring light intensity. *See* PAR METER.

**photomixotroph:** *See* MIXOTROPHIC CULTURES.

**photomorphogenesis:** The effects of light on plant growth and development. Light effects include day length, light intensity, and the effects of wavelength (Singhal et al., 1999). *See* ARTIFICIAL LIGHT; BLUE LIGHT RESPONSE; GROWTH ROOM LIGHTING; PHYTOCHROME.

**photon:** A quantum of electromagnetic energy; a discrete stable particle with zero mass and no electric charge. *See* CHLOROPHYLL FLUORESCENCE; ELECTROMAGNETIC SPECTRUM.

**photon flux density:** Photon irradiance expressed in moles per square meter per second. *See* PHOTOSYNTHETICALLY ACTIVE RADIATION.

**photoperiod:** The length of the light period in alternating light-dark sequences. The photoperiod influences various physiological responses in plants (Singhal et al., 1999). *See* LONG-DAY PLANT; PHYTOCHROME; SHORT-DAY PLANTS.

**photorespiration:** Light-dependent respiration occurring in the glyoxysomes in which photosynthetic efficiency is reduced by almost 50 percent (Ridge, 2002).

**photosynthetic pigments:** Pigments involved in trapping light during photosynthesis and the site of light–energy transformation. The major pigment involved is chlorophyll a. Accessory pigments that trap light and pass energy to chlorophyll a include chlorophyll b, carotenoids, and xanthophylls (Raghavendra, 1998; Rao, 1999; Ridge, 2002; Taiz and Zeigler, 2002). *See* photosynthesis.

**photosynthesis:** A two-stage transformation of light energy into chemical energy in photosynthesizing organisms. The first stage involves photolysis of water with the formation of adenosine triphosphate (ATP) and reduced nicotinamide adenine dinucleotide phosphate (NADPH). The second phase involves exploiting these energy-rich compounds in the dark reaction in which carbon is built into carbohydrates (Raghavendra, 1998; Rao, 1999; Ridge, 2002; Taiz and Zeigler, 2002). The photosynthetic pigments absorb light in the region 400-700 nm (Ridge, 2002). *See* ARTIFICIAL LIGHT; GROWTH ROOM LIGHTING; PHOTOSYNTHETICALLY ACTIVE RADIATION; PHOTOSYNTHETIC PIGMENTS.

**photosynthetically active radiation:** Light absorbed by the photosynthetic pigments in the range 400-700 nm (Taiz and Zeigler, 2002). *See* ELECTROMAGNETIC SPECTRUM; PHOTOSYNTHESIS.

**photosynthetic photon flux density:** PPFD; photosynthetically active radiation, i.e., light in the region 400-700 nm, expressed as $moles \cdot m^{-2} \cdot sec^{-1}$. *See* PAR METER.

**phototroph:** An organism synthesizing organic compounds with - energy derived directly from sunlight. *See* HETEROTROPHIC GROWTH; MIXOTROPHIC CULTURES.

**pH paper:** Paper that indicates, by color change, whether a solution is acid or alkaline (litmus paper) or, by graded color changes, pH ranges.

**phthalates:** Compounds used to make flexible plastic films. They can be toxic to some plant species. *See* PLASTICIZER; POLYVINYL CHLORIDE.

**phyllotaxy:** Genetically determined arrangement of leaves on a stem (Esau, 1977; Howell, 1998; Graham et al., 2003).

**physical mutagen:** Use of electromagnetic radiation to induce mutation (Micke and Donini, 1993). *See* CHEMICAL MUTAGENS; MUTATION BREEDING.

**physiological age:** The temporal age of a plant can be calculated from germination, but with plants growing at different rates, speed of development may vary, leading to physiological status being different at different temporal stages, such stages being defined as physiological ages for comparison (Howell, 1998). *See* PHASE CHANGE.

**phytic acid:** A common phosphate storage compound in the form of calcium-magnesium salts of hexaphosphoric esters (Metzler, 2003).

**phytoalexins:** Diverse secondary compounds induced in plants on pathogen infection. They accumulate locally and inhibit pathogen infection (Strange, 2003).

**phytochrome:** A family of plant-growth-regulating chromophore proteins that absorb light in the red, far-red, and blue regions. Phtyochrome exists in two forms: Pr in the dark (red-light absorbing) and Pfr in the light (far-red absorbing) (Taiz and Zeigler, 2002). *See* GROWTH ROOM LIGHTING.

**phytogel:** A commercial gelling agent. *See* GELLING AGENT.

**phytohormone:** A plant hormone, plant growth regulator, or plant bioregulator. *See* PLANT GROWTH REGULATOR.

**phytol:** A chlorophyll component comprising a hydrophobic long-chain alcohol with four isoprene units (Metzler, 2003).

**phytosanitary certificate:** An official certificate accompanying international plant shipments that states the material has been tested for pathogens as specified in the import license and is free of other contamination (Cassells and Doyle, 2005). *See* IMPORT LICENSE; PATHOGEN INDEXING; PLANT HEALTH CERTIFICATION.

**phytotron:** Temperature- and humidity-controlled high-light environmental chamber for autotrophic plant growth under strictly defined conditions. The main difference between a phytotron and a plant growth room is that the light intensity in a phytotron (1,000 to 2,000 $\mu$moles·m$^{-2}$·sec$^{-1}$) is 10 to 20 times greater than the light in a growth room (10 to 100 $\mu$moles·m$^{-2}$·sec$^{-1}$) (George, 1993). *See* AUTOTROPHIC CULTURE.

**P$_i$:** Parental generation initiating a cross in a breeding experiment.

**picloram:** An auxinlike synthetic growth regulator that has herbicidal activity (Basra, 2000) but can induce rapid callus proliferation when added to tissue culture medium (George, 1993). Picloram-tolerant cell lines can be selected (Chaleff and Parson, 1978). *See* AUXINS.

**pigment:** A colored compound absorbing light of specific wavelengths, e.g., chlorophyll and compounds coloring petals and fruit (Hendry, 1993).

**pinocytosis:** The uptake of molecules by clathrin-induced invagination of the plasmalemma. Pinocytotic vesicles so produced are 0.05-0.25 nm in diameter (Alberts et al., 2002).

**pipette:** A calibrated device for the transfer of either fixed or variable volumes of liquid. Pasteur pipettes are simple drawn-out lengths of glass tubing with a rubber bulb.

**pistil:** A structure made up of ovary, style, and stigma, which together form the main female floral organ (Esau, 1977).

**pith:** Tissue occupying the center of the stem and the inner zone of the root; vascular tissue made up of parenchyma cells (Esau, 1977).

**plagiotropic:** Horizontal or prostrate growth habit (Hartmann et al., 2001). *See* ORTHOTROPIC.

**plant adventitious regeneration:** *See* ADVENTITIOUS REGENERATION.

**plant anatomy:** Plant structure. It is important to understand cell specialization within the plant tissues to appreciate which cells are living and capable of dedifferentiation to undergo subsequent division and organogenesis (Trigiano and Gray, 2005; Figure 23). *See* COMPETENCE; MERISTEM; PARENCHYMA.

**plant bioregulator:** *See* PLANT GROWTH REGULATOR.

**plant biotechnology:** For some, biotechnology is based on the application of recombinant DNA technology; others broaden the definition to include applications of plant cell, tissue, and organ culture. *See* BIOTECHNOLOGY.

**Plant Breeders' Rights:** A scheme administered by the International Union for the Protection of New Varieties of Plants (UPOV). As with patents, the aim is to encourage invention, here the production of new varieties, by allowing the breeder a period of time to recover costs and make profits without competition. In 1991 plant breeders' rights were strengthened with the introduction of the *essential derivation* principle, which prevents third parties using protected varieties (Berman, 2002). *See* INTERNATIONAL UNION FOR THE PROTECTION OF NEW VARIETIES OF PLANTS; MUTATION BREEDING.

**plant cell wall:** The coat on the plant cell external to the plasmalemma. It is constructed from cellulose and other polymers and contains some proteins. The primary cell wall is made up of cellulose, hemicellulose, and pectic materials. When cell extension occurs, a secondary wall is formed of the same materials. In some tissues (e.g., xylem, sclerenchyma), the cell walls undergo secondary thickening with lignin or suberin. In addition to protecting the cell against pests and predators, the cell wall provides structural support for the plant. Cell turgidity, the pressure of the hydrated protoplasts against the cell

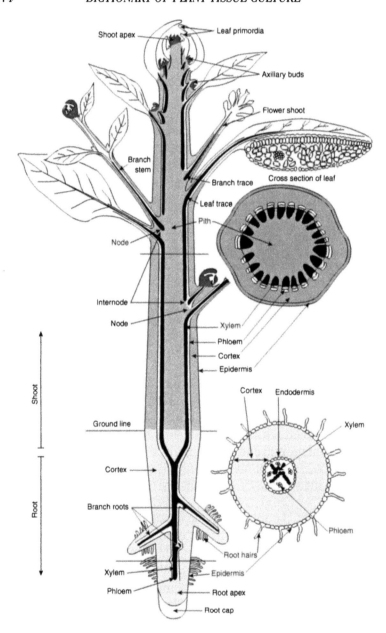

FIGURE 23. A representation of the main plant tissues and organs (Weier et al., 1974). *See* CAMBIUM; COMPETENCE; EXPLANT; MERISTEM; PARENCHYMA.

wall, maintains the rigidity of soft tissues such as leaves (Brett and Waldron, 1989; Carpita et al., 2001). *See* APOPLASTIC MOVEMENT PROTOPLAST ISOLATION.

**plant-growth-promoting rhizobacteria:** PGPR; root-zone inhabiting bacteria that promote plant growth and may induce systemic resistance to pathogens (Ryder et al., 1995). They are used as inoculants for microplants (Cordier et al., 2000). *See* BIOTIZATION; MYCORRHIZAL FUNGI.

**plant growth regulator:** Plant growth substance, including plant hormones and synthetic plant-growth-regulating compounds. Plant hormones include abscissic acid, auxin, brassinosteroids, cytokinins, ethylene, jasmonates, and gibberellin and their natural inhibitors and antagonists. Plant growth regulators include synthetic compounds with activities resembling their natural analogs and including antagonists and inhibitors. Many processes are controlled not by single plant growth regulators but by interactions between plant growth regulators and other compounds and systems (Moore, 1989; Roberts and Tucker, 1999; Hopkins and Hunter, 2002; Taiz and Zeigler, 2002; Figures 24-26; Table 3). *See* PLANT TISSUE CULTURE MEDIA.

**plant health certification:** An official scheme regulating the testing of plants for pathogens. Plant health certification is organized by the United Nations Food and Agriculture Organization (FAO) through a series of regional organizations, e.g., the North American Plant Protection Organization (NAPPO) and the European and Mediterranean Plant Protection Organization (EPPO), which agree upon the test procedures. Certification will be accepted only if official test methods are used (Krczal, 1998; Mink, 1998; Van der Linde, 2000). *See* IMPORT LICENSE; PHYTOSANITARY CERTIFICATE.

**plantlet:** *See* MICROPLANT.

**plant propagation:** Conventionally, plants are propagated by seed or from vegetative parts (Hartmann et al., 2001). Micropropagation is an alternative method of propagation based on plant tissue culture

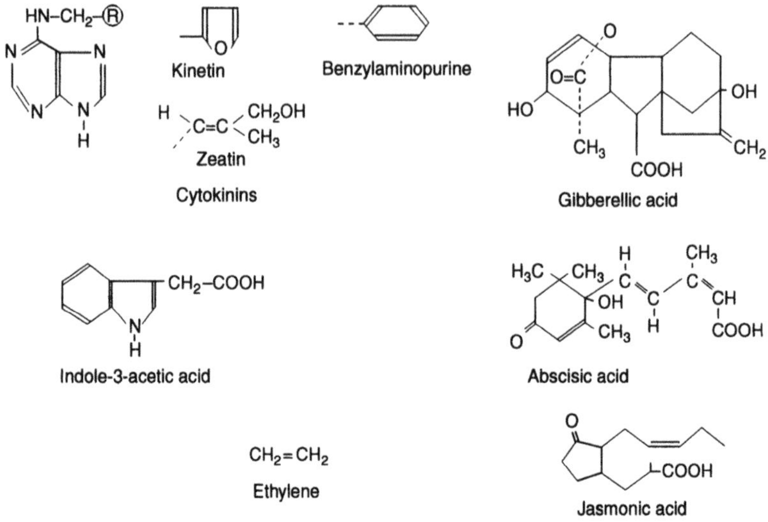

FIGURE 24. Structure representative of the main plant growth regulators (based on Marschner, 1994). Reprinted by permission. *See* PLANT GROWTH REGULATORS and Table 3.

methods (George, 1993, 1996). See MICROPROPAGATION; PLANT TISSUE CULTURE.

**plant nutrient:** An essential element for plant growth required in tissue. *See* MACRONUTRIENTS; MICRONUTRIENTS.

**plant patent:** Legal protection for plant breeders' intellectual property rights initially enacted in the United States and now being adopted in Europe. The principle of essential derivation is in accordance with International Union for the Protection of New Varieties of Plants (UPOV) regulations (Berman, 2002). *See* PLANT BREEDERS' RIGHTS.

**plant propagation:** *See* VEGETATIVE PROPAGATION.

**plant tissue culture:** The in vitro culture of plant cells, tissues, organs (George, 1993, 1996; Trigiano and Gray, 2000, 2005).

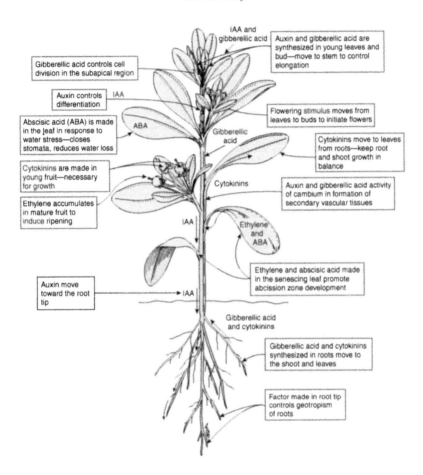

IAA and
gibberellic acid

Auxin and gibberellic acid are
synthesized in young leaves and
bud—move to stem to control
elongation

Gibberellic acid controls cell
division in the subapical region

Auxin controls
differentiation

IAA

Abscisic acid (ABA) is made
in the leaf in response to
water stress—closes
stomata, reduces water loss

ABA

Gibberellic
acid

Flowering stimulus moves from
leaves to buds to initiate flowers

Cytokinins move to leaves
from roots—keep root
and shoot growth in
balance

Cytokinins are made in
young fruit—necessary
for growth

Cytokinins

Auxin and gibberellic acid activity
of cambium in formation of
secondary vascular tissues

Ethylene accumulates
in mature fruit to
induce ripening

IAA

Ethylene
and
ABA

Ethylene and abscisic acid made
in the senescing leaf promote
abcission zone development

Auxin move
toward the root
tip

IAA

Gibberellic acid
and cytokinins

Gibberellic acid and cytokinins
synthesized in roots move to
the shoot and leaves

Factor made in root tip
controls geotropism
of roots

FIGURE 25. The main sites of synthesis and patterns of movement of plant growth regulators (Weier et al., 1974). *See* AUXIN (TRANSPORT) INHIBITORS; ETHYLENE INHIBITORS.

**plant tissue culture media:** Differs from in vivo plant mineral nutrient media in having a carbon growth source, organic supplements, and plant growth regulators, usually combinations of auxins and cytokinins. The Murashige and Skoog (1962) medium is widely used, but dedicated media exist for different plant types (e.g., woody plants and orchids) and for specific cell or tissue

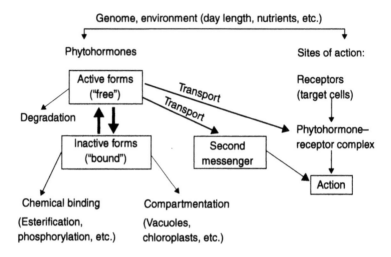

FIGURE 26. Factors influence the tissue concentration ('free' forms') of plant growth regulators (syn. phytohormones) (from Marschner, 1994) and their modes of action via secondary messengers or phytohormone receptors. Reprinted by permission. *See* AUXIN INHIBITORS; ETHYLENE RECEPTORS; PLANT GROWTH REGULATORS.

types (e.g., protoplasts and anther culture) (George, 1993, 1996; Caponetti et al., 2005; Table 4). *See* MICROPROPAGATION; PROTOPLAST CULTURE.

**plant transformation:** *See* GENETIC ENGINEERING.

**plant viruses:** Viruses infecting plants, most of which contain RNA as the genetic material (Hull, 2002). VIRUS ELIMINATION; VIRUS FREE; VIRUS TESTED.

**plasmalemma:** Plasma membrane surrounding a plant protoplast (Alberts et al., 2002; Taiz and Zeigler, 2002).

**plasmid:** A small, circular DNA molecule found in bacteria and fungi that replicates independently in the proliferating host cell (Russell, 2002). *See* AGROBACTERIUM RHIZOGENES; AGROBACTERIUM TUMEFACIENS.

TABLE 3. Plant growth regulators commonly used in tissue culture media.

| Type | Abbreviation | Chemical name |
|------|-------------|---------------|
| Abscisic acid | ABA | (2$Z$,4$E$)-5-[(1$S$)-1-hydroxy-2,6,6-trimethyl-4-oxo-2-cyclohexen-1-yl]-3-methyl-2,4-pentadienoic acid |
| Auxin | 2,4-D | 2,4-dichlorophenoxyacetic acid |
| | 2,4,5-T | 2,4,5-trichlorophenoxyacetic acid |
| | Dicamba | 2-methoxy-3,6-dichlorobenzoic acid |
| | IAA | Indole-3-acetic acid |
| | IBA | Indole3-butyric acid |
| | MCPA | 2-methyl-4-chlorophenoxyacetic acid |
| | NAA | 1-naphthylacetic acid |
| | NOA | 2-naphthyloxyacetic acid |
| | Picloram | 4-amino-2,5.5-trochloropicolinic acid |
| Auxin inhibitors | TIBA | 2,3,5-triiodobenzoic acid |
| Cytokinin | BA/BAP | 6-benzylaminopurine |
| | 2iP (IPA) | [$N^6$-(2-isopentyl)adenine] |
| | Kinetin | 6-furfurylaminopurine |
| | Thidiazuron (TDZ) | 1-phenyl-3-(1,2,3-thiadiazol-5-yl)urea |
| | Zeatin | 4-hydroxy-3-methyl-trans-2-butenylaminopurine |
| Gibberellin | GA$_3$ | (3$S$,3a$S$,4$S$,4a$S$,7$S$,9a$R$,9b$R$,12$S$)-7,12-dihydroxy-3-methyl-6-methylene-2-oxoperhydro-4a,7-methano-9b,3-propeno[1,2-$b$]furan-4-carboxylic acid |
| Antigibberellin | Ancymidol | (RS)-$\alpha$-cyclopropyl-4-methoxy-$\alpha$-(pyrimidin-5-yl)benzyl alcohol |
| | Chlormequat (CCC) | 2-chloroethyltrimethylammonium |
| | Daminozide | $N$-dimethylaminosuccinamic acid |

*Source:* Based on Gaba, V.P. (2005). Plant growth regulators in tissue culture and development. In Trigiano, R.N. and Gray, D.J. (Eds.), *Plant Development and Biotechnology* (pp. 87-100). Boca Raton, FL: CRC Press.

TABLE 4. Components of plant tissue culture media.

| Category and components | Comments |
|---|---|
| Water | High-grade water free of microorganisms, minerals, and organic residues is required. *See* WATER QUALITY |
| Mineral nutrients | |
|   Macronutrients<br>  Micronutrients<br>  Beneficial minerals | See Table 1. Reagent-grade chemicals may contain inorganic contaminants. *See* MINERAL NUTRITION |
| Organic substances | |
|   Myo-inositol | Present in Murashige and Skoog (1962) medium and in most media |
|   Amino acids and amides | Glycine is present in Murashige and Skoog (1962) medium and in many media; asparagines, L-glutamine, serine, and proline are sometimes used (Beyl, 2005) |
|   Vitamins | Nicotinic acid, pyridoxine HCl, and thiamine HCl are present in Murashige and Skoog (1962) medium and in most media; ascorbic acid, biotin, folic acid, pantothenic acid, para-aminobenzoic acid, and tocopherol are sometimes used (Beyl, 2005) |
| Carbon source | |
|   Sucrose | Present in Murashige and Skoog (1962) medium and in most media. At high concentration may inhibit Rubisco, the key enzyme in photosynthesis. *See* RIBULOSE BISPHOSPHATE CARBOXYLASE |
|   Other sugars | Glucose, fructose, maltose, and sorbitol are occasionally used (Beyl, 2005) |
| Polyamines | |
|   Putrescine, spermidine | These and other polyamines are incorporated into embryogenic and rooting media (George, 1993, 1996) |
| Undefined supplements | |
|   Complex mixtures | Coconut milk was historically widely used as a media supplement but is variable in composition, yielding variable results. Also used occasionally are fruit extracts. *See* COCONUT MILK |

TABLE 4 *(continued)*

| Category and components | Comments |
| --- | --- |
| Protein hydrolysates | Casein hydrolysate, peptone, tryptone, malt, and yeast extracts are sometimes used as sources of amino acids and of vitamins, etc. They are variable in composition. Casein hydrolysate is also used to encourage the expression of bacterial contamination. *See* MENARD'S MEDIUM |
| Other | |
| Activated charcoal | Of variable purity, binds mainly aromatic compounds; darkening of medium may also promote light-sensitive reactions; used occasionally in rooting media. *See* ACTIVATED CHARCOAL |
| Antioxidants | Ascorbic acid, citric acid, and thiourea are added to prevent polymerization of phenols. *See* OXIDATIVE STRESS |
| Polyvinylpyrrolidone (PVP) | Binds phenols |
| pH | pH affects the solubility of mineral nutrients and the setting of agar. Some effects of bacterial contaminants may be mediated via changes in medium pH. pH indicators are sometimes aged to media (George, 1993). *See* pH |
| Tissue support | |
| Gelling agent | Agar is the most commonly used tissue support; sometimes clear gelling agents (e.g., Gelrite) are used to facilitate examination of the cultures for contamination. Most gelling agents are natural products and vary in quality (Cassells and Collins, 2000). *See* GELLING AGENT |
| Raft | Paper and plastic rafts are used as an alternative to gelling agents. *See* PAPER RAFT METHOD; RAFT; TEMPORARY IMMERSION |

*Source:* Beyl, C.A. (2005). Getting started with tissue culture: Media preparation, sterile technique, and laboratory equipment. In Trigiano, R.N. and Gray, D.J. (Eds.), *Plant Development and Biotechnology* (pp. 19-37). Boca Raton, FL: CRC Press.

**plasmodesma, pl. plasmodesmata:** Narrow passages, lined by the plasmalemma and connected to the endoplasmic reticulum, that pass through cell walls to connect adjacent cells and through which substances such as RNA, DNA, and viruses can pass (Esau, 1977; Hull, 2002). *See* PHLOEM-RESTRICTED MICROORGANISMS.

**plasmolysis:** The separation of the plasmalemma from the cell wall as a result of a loss of water from the protoplast (Ridge, 2002). *See* PROTOPLAST ISOLATION.

**plasmolytic vesicle:** Vesicle formed by invagination of the plasmalemma and often occurring during the removal of the cell wall during the preparation of protoplasts (Warren, 1991). *See* PROTOPLAST ISOLATION.

**plastic containers:** *See* GASEOUS PERMEABILITY OF PLASTICS; PLASTIC FILMS.

**plastic culture vessels:** *See* CULTURE VESSELS.

**plastic films:** Films made from natural or synthetic polymers, which may vary in thickness and flexibility (Osborn and Jenkins, 1992). Plastic materials are also used to produce tissue culture vessels (George, 1993). *See* GASEOUS PERMEABILITY OF PLASTICS.

**plasticity:** Capacity of a plant to modify the form of newly developing tissues to an environmental condition (Howell, 1998).

**plasticizer:** A substance that when added to a material, usually a plastic, makes it flexible, resilient, and easier to handle and may modify its gaseous permeability. Plasticizers are esters, e.g., adipates and phthalates (Osborn and Jenkins, 1992). *See* GASEOUS PERMEABILITY OF PLASTICS; PHTHALATES.

**plastics:** Materials made from synthetic or natural polymers. Plastics differ in transparency, permeability to gases, heat stability, and resistance to sterilizing agents. They are used to make tissue culture vessels and vessel lids. See PLASTIC FILMS; TISSUE CULTURE VESSELS.

**plastid:** Interconvertible cytoplasmic organelle often containing pigments and found in most plant cells. Plastids vary in size and structure from small proplastids to large chromoplasts, chloroplasts, and storage plastids, e.g., amyloplasts and elaioplasts (Graham et al., 2003).

**plate:** The distribution of a thin layer of, e.g., protoplasts or cells or microorganisms onto a culture medium surface.

**platform shaker:** *See* SHAKER.

**plating efficiency:** Percentage of viable cell colonies developing on an inoculated plate.

**pleiotropy:** The effect of a single gene on a number of characteristics (Allard, 1999; Russell, 2002). Microplants may show juvenility-related pleiotropic disease resistance (Cassells et al., 1991).

**plerome:** Apical meristem tissue giving rise to all primary tissues internal to the cortex (Lyndon et al., 1998).

**ploidy:** The number of complete sets of chromosomes present in a cell, which is denoted by the letter $n$ (Allard, 1999).

**plumule:** The embryonic axis above the cotyledonary node (Esau, 1977).

**point mutation:** Mutation affecting a single base in a DNA sequence. Point mutation is induced at high frequency by chemical mutagens. *See* CHEMICAL MUTAGENS; PHYSICAL MUTAGEN.

**polarity:** Morphological and physiological differences between the poles of a structure (e.g., cells, whole plants) that appear to be set at the formation of the zygote.

**polarizing microscope:** A microscope employing illumination of the specimen with polarized light. Oriented structures in the specimen (e.g., crystals, starch grains, cell walls) can cause rotation of the light, leading to birefringence of the specimen, which appears light on a dark background. The rotation of light can be measured to yield information on the specimen's structure (Gahan, 1984; Murphy, 2001).

**polar nuclei:** The two haploid nuclei derived from the megaspore and found in the center of the embryo sac. They can fuse to form a diploid nucleus prior to fusion with the male gamete to form the triploid primary endosperm nucleus (Graham et al., 2003).

**pollen culture:** Pollen grains cultured in vitro to generate callus from which haploid microplants can be derived (Nitsch, 1977; Reed, 2005).

**pollen grain:** Haploid microspores of seed plants derived by meiosis (Graham et al., 2003). *See* POLLEN CULTURE.

**pollen sac:** A sac within the anther, in which microspores are produced (Graham et al., 2003).

**pollen tube:** The tube growing toward the oocyte and carrying male gametes derived from pollen grain (Graham et al., 2003). *See* INCOMPATIBILITY; IN VITRO POLLINATION.

**pollinate:** Transfer of pollen from anther to stigma (Russell, 2002; Mauseth, 2003).

**polyacrylamide gel electrophoresis:** PAGE; a technique for separating macromolecules based on size and shape. The gel pore size may be varied to produce different molecular sieving effects for separating proteins of different sizes. By controlling the percentage (from 3 to 30 percent), precise pore sizes can be obtained, usually from 5 to 2,000 kDa. Polyacrylamide gels can be cast in a single percentage or with varying gradients. Polyacrylamide gels offer greater flexibility and more sharply defined banding than agarose gels (Hames and Rickwood, 2001). *See* GEL ELECTROPHORESIS.

**polyamine:** A compound with two or more amino groups (Slocum and Flores, 1991). Polyamines include putrescine, spermidine, and spermine, though other related and/or less common members are known. Polyamines are small, ubiquitous organic polycations that have been implicated in a wide variety of physiological functions, including protein translation, membrane stabilization, and cell proliferation. Polyamines protect against abiotic and biotic stress (Walters, 2003; Kasukabe et al., 2004) and they have been shown to promote plant growth (Nassar et al., 2003). Polyamines have been associated with organogenesis and embryogenesis in vitro (George, 1993; Gaba, 2005).

**polycarbonate:** Plastic material used for culture vessels. *See* GASEOUS PERMEABILITY OF PLASTICS.

**polyembryony:** The formation of more than one embryo, usually from the zygote though occasionally from somatic tissue (Bajaj, 1995).

**polyethylene glycol:** PEG; water-soluble polymer used to solubilize proteins and as a cell fusogen (Warren, 1991). PEG can be used as an osmoticum in tissue culture.

**polygenes:** A number of nonallelic genes all of which affect the same character. They are additive in their effects (Allard, 1999).

**polymerase chain reaction:** PCR; a technique involving repeated cycles of strand separation (denaturation), annealing with primer, and extension with DNA polymerase. PCR is used to amplify a target DNA sequence by more than a millionfold, making the product visible after staining on gels (Candress et al., 1998; McPherson et al., 2000). *See* GENETIC ENGINEERING; PATHOGEN INDEXING.

**polymixin:** An antibiotic produced by *Bacillus polymyxa,* used clinically against gram-negative organisms and some fungi (Walsh, 2003). *See* ANTIBIOTICS.

**polynucleotide:** A sequence of nucleotides covalently linked by a phosphate group through the 3' position of one pentose molecule to the 5' position of the next (Alberts et al., 2002). *See* DEOXYRIBONUCLEIC ACID; RIBOSE NUCLEIC ACID.

**polypeptide:** A sequence of amino acids linked in a chain by a covalent bond between the carboxyl group of one amino acid and the amino group of the next to form the primary structure of a protein molecule (Alberts et al., 2002).

**polyphenols:** Polymers of phenols (Hemingway and Laks, 1992). *See* PHENOLS.

**polyploidy:** Presence of three of more complete sets of chromosomes per somatic cell nucleus (Allard, 1999). *See* EUPLOID.

**polyribosome:** Syn. polysome; a sequence of cytosolic ribosomes linked by messenger RNA (Alberts et al., 2002).

**polysaccharide:** A linear or branched chain of monosaccharide molecules linked through glycosidic bonds (Bryant et al., 1999).

**polysomy:** Several copies of a particular chromosome in a cell or organism (Allard, 1999).

**polyurethane foam:** A common foam used in furnishings and as a tissue support in plant tissue culture. Caution: most polyurethane foams contain fire retardants, which are toxic to plants (Cassells, 2000b). *See* TISSUE SUPPORTS.

**polyvinyl alcohol:** PVA; PVA comes in variable molecular weights and is added as a tissue stabilizer prior to freezing (Gahan,1984). *See* CRYOPRESERVATION.

**polyvinyl chloride:** Clear plastic with relatively high permeability to water vapor, carbon dioxide, oxygen, and ethylene depending on film thickness. It is used for tissue culture vessel lids (Cassells and Roche, 1994; Cassells and Walsh, 1994). *See* PLASTIC FILMS; TISSUE CULTURE VESSELS.

**polyvinylpyrrolidone:** PVP; an antioxidant that comes in variable molecular weights and is added to prevent browning of explanted tissues. *See* MENARD'S MEDIUM.

**population density:** Cell number per unit of medium area or density (Hunter, 1993).

**population doubling time:** *See* DOUBLING TIME.

**porphyrins:** Compounds found in pigments. Porphyrins contain four pyrrole groups linked into a ring by methane groups between their α-carbons, e.g., chlorophylls, cytochromes (Hendry, 1993; Taiz and Zeigler, 2002).

**postzygotic incompatibility:** A condition where zygotes from distant or incompatible crosses fail to develop for nutritional reasons (de Nettancourt, 1993). *See* EMBRYO RESCUE; PREZYGOTIC INCOMPATIBILITY.

**potassium:** K; present in high concentrations in cells and important as a preferred cation in tissues. Many enzymes require potassium for activity (Metzler, 2001). *See* MACRONUTRIENTS; PLANT TISSUE CULTURE MEDIA.

**potassium hydroxide:** Used to adjust the pH of culture media and as a solvent for auxins.

**potassium nitrate:** Main nitrogen source in culture media for somatic embryogenesis (Dodds and Roberts, 1995). *See* PLANT TISSUE CULTURE MEDIA.

**potato extract:** A common addition to culture media for anther and monocot cultures.

**PPFD:** *See* PHOTOSYNTHETIC PHOTON FLUX DENSITY.

**ppm:** Parts per million.

**precipitate:** Solid material coming out of solution.

**precipitin test:** The identification of antigens by causing them to precipitate in the presence of antibodies (Dijkstra and de Jeger, 1998). *See* ANTIBODY; SEROLOGICAL TESTS.

**precocious germination:** Undesirable early germination encountered with somatic embryos, which may lack the dormancy required for use in artificial seed (Redenbaugh, 1993; Bajaj, 1995). *See* ARTIFICIAL SEED.

**precondition:** Enhancement of culture growth by prior pretreatment of the stock plant (George, 1993) or culture medium (Warren, 1991).

**premix:** Commercial mixtures of dry, preweighed ingredients ready for dissolving in solution when required. *See* PLANT TISSUE CULTURE MEDIA.

**preserve:** Maintain in good condition.

**pressure cooker:** Small-scale autoclave useful for sterilizing small quantities of medium or instruments (George, 1993).

**pressure potential:** *See* TURGOR PRESSURE.

**pretransplant:** Stage 3 in micropropagation in which rooting and hardening for transfer to soil (Stage 4) occur (George, 1993). *See* MICROPROPAGATION.

**prezygotic incompatibility:** Barriers to fertilization found in wide or incompatible crosses (de Nettancourt, 1993). *See* IN VITRO POLLINATION; POSTZYGOTIC INCOMPATIBILITY.

**primary cell wall:** The initial cell wall formed after cell division (Brett and Waldron, 1989; Chapman and Waldron, 1996; Carpita et al., 2001).

**primary culture:** A culture that has yet to be subcultured.

**primary explants:** Explants from the stock plant used to establish an in vitro culture.

**primary growth:** Increase in size as the result of cell division in the primary meristems and subsequent cell enlargement (Lyndon et al., 1998).

**primary xylem:** Derived from the procambium in the primary plant body (Esau, 1977; Tyree and Zimmermann, 2002).

**primordium:** Immature part of a plant programmed to develop into a specific cell, tissue, or organ (Howell, 1998).

**proembryo:** Initial embryo prior to the globular form (Bhojwani and Soh, 2001).

**progeny:** Offspring derived by either sexual or asexual reproduction (Allard, 1999).

**programmed cell death:** *See* APOPTOSIS.

**projectile bombardment:** *See* PARTICLE BOMBARDMENT; PARTICLE GUN.

**prokaryote:** A small cell lacking a nuclear membrane and organelles. Prokaryotes are the possible origin of the mitochondria and chloroplasts of the eukaryotic cell. They include bacteria, Cyanophyceae, and Chlorophyceae (Prescott et al., 2001). *See* ANTIBIOTICS; EUKARYOTES.

**promeristem:** Embryonic meristem containing cells already determined for organogenesis (Esau, 1977; Howell, 1998).

**propagation:** *See* PLANT PROPAGATION.

**propagule:** Part of an organism used for multiplication.

**propiconazole:** A fungicide that inhibits ergosterol biosynthesis in fungi, so blocking cell wall formation (Strange, 2003). *See* ANTIBIOTICS.

**propylene oxide:** Colorless volatile liquid used to sterilize plant material and soil. As a solvent it is used for histological embedding resin (Gahan, 1984). *See* SURFACE STERILANTS.

**protease:** An enzyme that splits the peptide bonds between adjacent amino acids in a polypeptide chain (Metzler, 2001).

**protein:** Sequence of the 20 basic, commonly occurring amino acids joined by covalent peptide linkages to form a polypeptide chain, which is folded into a secondary and biologically active tertiary structure held together by noncovalent and covalent bonds (Branden and Tooze, 1998; Metzler, 2001). *See* AMINO ACIDS.

**protein chips:** *See* PROTEIN MICROARRAYS.

**protein digest:** Protein hydrolysate containing amino acids, peptides, and short-chain polypeptide chains.

**protein microarrays:** An ordered array of proteins with specific binding characteristics immobilized on a solid substrate, e.g., a silicon chip. The proteins can include antibodies and enzymes (Schena, 2004). *See* DNA MICROARRAYS.

**protein synthesis:** Syn. translation; the formation of a polypeptide chain on the ribosome according to instructions from a specific mRNA and with amino acids transported to the site by tRNA. The ribosomes may be situated in the cytosol for the production of proteins needed for, e.g., mitochondria, chloroplasts, and glyoxisomes. Alternatively, the ribosomes may attach to the endoplasmic reticulum to form proteins to be either secreted or sent to vacuoles (Alberts et al., 2002).

**proteomics:** The study of the structure and function of proteins (Liebler, 2002). *See* GENOMICS; METABOLOMICS.

**Protista:** Kingdom containing bacteria, protozoa, algae, fungi, and slime molds.

**protoclone:** Clone derived from a single protoplast or product of protoplast fusion.

**protocol:** Recipe for achieving a sequence of activities or manipulations.

**protoderm:** The outer layer of a meristem, from which the epidermis and associated subepidermal layers are derived (Lyndon et al., 1998).

**protomeristem:** *See* PROMERISTEM.

**protophloem:** First-formed primary phloem (Esau, 1977).

**protoplasm:** Protoplast contents between the nucleus and the plasmalemma (Graham et al., 2003).

**protoplast:** A plant cell from which the cell wall has been removed (Warren, 1991).

**protoplast culture:** In vitro culture of protoplasts. Media modifications are generally required for protoplast culture, and nurse culture is frequently necessary (Warren, 1991). Ethylene and ethane release by the isolated protoplasts is an indicator of protoplast potential viability (Cassells and Tamma, 1985), and cell wall regeneration is the first indicator of the potential for cell division. *See* NURSE CULTURE.

**protoplast fusion:** Fusion of two or more protoplasts aimed at the production of a somatic hybrid bypassing incompatibility barriers (Bajaj, 1996a; van Tuyl et al., 2002; Veilleux et al., 2005). *See* CHEMICAL FUSION; ELECTROFUSION; INCOMPATIBILITY; PROTOPLAST ISOLATION.

**protoplast isolation:** A procedure involving cell plasmolysis and cell wall enzymatic digestion. The cells are usually plasmolyzed in hypertonic mannitol and digested with a cocktail of enzymes including cellulases and pectinases (Warren, 1991). *See* PROTOPLAST CULTURE; PROTOPLAST FUSION; PROTOPLAST VIABILITY.

**protoplast viability:** Usually determined by the use of vital dyes such as fluorescein diacetate (Gahan, 1984, 1989). *See* VITAL DYE.

**protoxylem:** The first-formed primary xylem (Esau, 1977; Tyree and Zimmermann, 2002).

**proximal:** Zone of an organ nearest to its point of attachment.

**Pteridophyta:** Vascular plants that are not seed bearing and show heteromorphic alternation of generations, with the gametophyte often dependent on the sporophyte for its nutrition; includes Psylopsida, Lycopsida, Sphenopsida, and Pteropsida (Graham et al., 2003; Mauseth, 2003).

**pubescent:** Describes organs or plants covered with either hair or trichomes (Esau, 1977).

**pure culture:** An axenic culture free of all other organisms. This status is difficult to confirm for plant tissue cultures (Cassells and Doyle, 2004). *See* AXENIC CULTURE.

**pure line:** Succession of generations breeding true, producing genotypically identical offspring when either selfed or crossed between themselves; hence, individuals are considered to be homozygous (Allard, 1999).

**purine:** A nitrogen-containing organic base with a double ring structure: adenine and guanine (Alberts et al., 2002). *See* DEOXYRIBONUCLEIC ACID; PYRIMIDINE; RIBOSE NUCLEIC ACID.

**PVC film:** *See* PLASTICS; POLYVINYL CHLORIDE.

**PVP:** *See* POLYVINYLPYRROLIDONE.

**pyrethrin:** A family of natural insecticides.

**pyrex:** Highly stable borosilicate glass.

**pyridoxine:** Vitamin B6; commonly included in plant tissue culture media (George, 1993). *See* PLANT TISSUE CULTURE MEDIA.

**pyrimidine:** A nitrogen-containing organic base with a single ring structure: cytosine, uracil, and thymine (Alberts et al., 2002). *See* DEOXYRIBONUCLEIC ACID; PURINE; RIBOSE NUCLEIC ACID.

**Q banding:** A fluorescent staining technique for chromosomes that produces specific banding patterns for each pair of homologous chromosomes (Fukui and Nakayama, 1996).

**quarantine:** The holding of imported material in isolation for a period to ensure freedom from diseases and pests (Schepin and Ermakov, 1991; Ebbels, 2003). *See* IMPORT LICENSE; PHYTOSANITARY CERTIFICATE; PLANT HEALTH CERTIFICATION.

doi:10.1300/5648_17

**R₁:** A regenerant from in vitro culture of either a shoot from a callus or a plantlet from a somatic embryo. R₂ and R₃ are designations for the second and third generations produced by crossing the R₁ and R₂ generations, respectively.

**radiant energy:** *See* ELECTROMAGNETIC SPECTRUM.

**radiant flux:** Radiant energy per unit of time (George, 1993).

**radiation:** Particles in wave form or rays of heat and light. *See* ELECTROMAGNETIC SPECTRUM.

**radiation damage:** Gamma and X-rays may induce mutation and also damage proteins, enzymes, and membranes (van Harten, 1998). *See* MUTATION.

**radicle:** Embryonic root emerging from the testa on germination (Esau, 1977).

**radioimmunoassay:** RIA; a method for the detection and quantification of substances using radioactively labeled antibody or antigen (Roitt et al., 2002). *See* SEROLOGICAL TESTS.

**radiolabeling:** Labeling molecules with radioactive atoms. The distribution of the radiolabel can be followed in time and space. The label can be detected using a liquid scintillation counter or autoradiographically by exposure of labeled biological material to photographic film (Gahan, 1972; Buncel and Jones, 1991). *See* GENETIC FINGERPRINTING.

**raffinose:** A common three-sugar compound consisting of fructose, galactose, and glucose units (Bryant et al., 1999).

**raft:** A support of tissues in culture that conducts nutrients to the tissue. It is an alternative to solidification of the medium (George, 1993). *See* GELLING AGENT; TISSUE SUPPORTS.

**ramet:** Individual belonging to a clone. *See* CLONE.

**random amplified polymorphic DNA:** RAPD; primer binding sites separated by defined lengths of intervening DNA used as markers (Edwards, 1998). *See* GENETIC FINGERPRINTING; POLYMERASE CHAIN REACTION.

**randomized block design:** An experimental design to account for variability in an experiment as a result of factors outside the experiment (Compton, 2005).

**RAPD:** *See* RANDOM AMPLIFIED POLYMORPHIC DNA.

**rate of propagation:** *See* MULTIPLICATION RATE.

**reaction norm:** The range of phenotypic responses to environmental influences by a particular genotype. *See* GENOTYPE; PHENOTYPE.

**reagent:** A chemical substance participating in a chemical reaction.

**recalcitrant:** Resistant to treatment, e.g., explants resistant to treatments to induce organogenesis.

**recalcitrant seed:** A seed failing to germinate despite a variety of pretreatments (Hartmann et al., 2001).

**recessive gene:** A gene that is expressed only in the homozygous allelic form and not normally in the heterozygous allele. It is suppressed by the dominant gene in a hybrid offspring (Allard, 1999). *See* ANTHER CULTURE; POLLEN CULTURE.

**recessive trait:** Opposite of dominant trait (Allard, 1999). *See* DOMINANT TRAIT.

**reciprocal cross:** A cross in which the source of male and female gametes is reversed (Allard, 1999). *See* CYTOPLASMIC MALE STERILITY.

**reciprocating shaker:** A variable-speed platform shaker moving backward and forward and used to agitate cultures in flasks.

**recombinant DNA:** Genetic material with novel gene sequences produced either by genetic engineering or through crossovers, chromosome assortment, or other natural mechanisms (Glick and Pasternak, 2003). *See* GENETIC ENGINEERING.

**recombinant DNA technology:** *See* GENETIC ENGINEERING.

**reconstructed cell:** Viable cell hybrid, cybrid, or cell transformed by genetic engineering.

**reculture:** Transfer either the whole or a portion of a culture to fresh culture medium. *See* SUBCULTURE.

**redetermination:** Reinitiation of determination in cells that appear to have lost this capacity (George, 1993).

**redifferentiation:** Alteration of the differentiated state of either a cell of a tissue from one form to another (George, 1993).

**red light:** Electromagnetic radiation of ~630 nm wavelength. *See* ELECTROMAGNETIC SPECTRUM; PHOTOMORPHOGENESIS; PHOTOSYNTHESIS.

**redox potential:** Tendency of a reducing agent to lose electrons or of an oxidizing agent to gain electrons (Favier et al., 1995).

**reducing agents:** Compounds with a strong affinity for oxygen; compounds that add an electron to an atom or ion (Favier et al., 1995).

**reducing sugars:** A sugar capable of reducing an oxidizing agent, e.g., all monosaccharides (Bryant et al., 1999).

**regenerate:** To renew or heal by new tissue growth; to reform or recreate. *See* ADVENTITIOUS REGENERATION.

**regeneration:** The process of regenerating. *See* ADVENTITIOUS REGENERATION.

**regenerative capacity:** The ability of cells or tissues to regenerate.

**reinvigoration:** Sometimes used as a synonym for rejuvenation. Reinvigoration refers to the recovery by woody plants of some juvenile characteristics following micrografting or serial subculture in vitro. The term "partial rejuvenation" is preferred (George, 1993). *See* MICROGRAFTING; REJUVENATION.

**rejuvenation:** Grafting of scions of the adult phases of woody plants to young rootstock (Hartmann et al., 2001); the invigoration or revival of cultures by micrografting or serial subculture in vitro (George, 1993). *See* REINVIGORATION.

**relative humidity:** Percentage of ambient water vapor with respect to a fully saturated atmosphere at a given temperature. As the relative humidity in vitro is close to saturation, cultures need to be adapted for outside relative humidities, e.g., in the field (George, 1993). *See* IN VITRO WEANING; MICROPLANT ESTABLISHMENT.

**Repelcote:** Trade name for dimethyl dichlorosilane coating; used on glassware and for tissue culture containers.

**replicate:** To repeat experiments or procedures; to duplicate.

**replication:** The process of duplication or repeating experiments; the process of cloning. The copying of DNA strands is semiconservative replication (Russell, 2002).

**reporter gene:** A gene incorporated into gene constructs to confirm the incorporation of the construct into the plant genome. Reporter genes may fluoresce or catalyze the formation of a pigment in the transformed cell, tissue, or transformed plant (Glick and Pasternak,

2003; Slater et al., 2003). *See* β-GLUCURONIDASE GENE; GENETIC ENGINEERING; GREEN FLUORESCENT PROTEIN; SELECTABLE MARKER.

**repression:** Lack of the expression of a gene leading to lack of a specific protein. The term generally refers to the blocking of the transcription of a gene by a repressor protein binding to a regulatory site or to the blocking of translation by binding of a repressor protein to mRNA (Allard, 1999; Russell, 2002).

**reproduce:** To produce an individual of the parental type.

**reproduction:** Process for forming individuals by sexual or asexual mechanisms. *See* SEXUAL REPRODUCTION; VEGETATIVE PROPAGATION.

**resistance:** Ability of an organism to resist change induced by either internal or external factors or disease. Resistance to such stresses can be induced (Allard, 1999; Strange, 2003). *See* ABIOTIC STRESS; BIOTIC STRESS.

**resistance transfer factor:** Plasmid present in some bacteria imparting antibiotic resistance to exposed animals (Russell, 2002; Walsh, 2003). *See* ANTIBIOTIC RESISTANCE.

**resistivity:** Reciprocal of conductivity. Resistivity is the degree of resistance to the flow of electric current or movement of particles. As a measure of water purity, high resistivity indicates high purity.

**respiration:** Gaseous exchange involving the uptake of oxygen and the release of carbon dioxide and water across a respiratory surface. Cell respiration involves the release and harnessing of energy from the breakdown of glucose for use in the cell (Taiz and Zeigler, 2002). *See* CITRIC ACID CYCLE.

**restriction endonuclease:** *See* RESTRICTION ENZYMES.

**restriction enzymes:** A group of enzymes that cleave DNA at specific sites (Glick and Pasternak, 2003). *See* DNA METHYLATION; METHYLATION-SENSITIVE RESTRICTION ENZYMES.

**restriction fragment:** Product of the digestion of DNA by restriction enzymes. Fragments are flanked by the recognition sites for the cleavage enzyme (Winfrey et al., 1997). *See* AMPLIFIED FRAGMENT LENGTH POLYMORPHISM TECHNIQUE; RESTRICTION FRAGMENT LENGTH POLYMORPHISM.

**restriction fragment length polymorphism:** RFLP; variation on the lengths of restriction fragments detected by gel electrophoresis; is used to identify organisms (Brettschneider, 1998). *See* GENETIC FINGERPRINTING.

**retrotransposon:** A transposon transcribed into RNA and subsequently reverse-transcribed to DNA that inserts into a new location in the genome (Capy, 1998; Russell, 2002). *See* TRANSPOSON.

**reverse genetics:** Process by which a specific gene is isolated, mutated at a specific site, and reintroduced into the genome, where its effects are studied. It is the opposite of mutant analysis (Russell, 2002; Slater et al., 2003). *See* MUTATION BREEDING.

**reverse osmosis:** Forcing a solution across a semipermeable membrane against the osmotic gradient in order to remove impurities.

**reversion to juvenility:** Reversion from adult to juvenile form inducible with plant growth compounds. Gibberellic acid causes rejuvenation in several woody angiosperms (Hartmann et al., 2001). *See* IN VITRO REJUVENATION.

**RFLP:** *See* RESTRICTION FRAGMENT LENGTH POLYMORPHISM.

**rhamnose:** Methylated pentose sugar. It is rarely found free but is present in glycosides (Bryant et al., 1999).

**rhizogenesis:** Root formation and growth (Davis and Hassig, 1994; Anderson et al., 1997; De Klerk, 2002). *See* MICROPROPAGATION.

**rhizome:** Normally horizontal, underground stem forming the primary shoot and bearing buds in the axils of reduced leaves. The rhizome is important in vegetative propagation and may also serve as a storage organ (Esau, 1977; Mauseth, 1988).

**rhodamine isothiocyanate:** A fluorescent dye used in the identification of protoplast fusion products. *See* PROTOPLAST FUSION.

**ribavirin:** Syn. Virazole; 1-β-D-ribofuranosyl-1,2,4-triazole-3-carboxamide; antiviral chemical developed for use against animal viruses. Ribavirin is the main antiviral chemical used in plant tissue culture (Cassells and Long, 1980a,b, 1982, 1983). *See* CHEMOTHERAPY.

**rib meristem:** Syn. file meristem; meristem producing files of cells. *See* Figure 8 (Esau, 1977).

**riboflavin:** Syns. vitamin B2, vitamin G; water-soluble compound essential to cell respiration. Riboflavin is involved in carbohydrate metabolism and perception of photoperiodic stimuli. It is occasionally added to culture medium (George, 1993, 1996). *See* PLANT TISSUE CULTURE MEDIA.

**1-β-D-ribofuranosyl-1,2,4-triazole-3-carboxamide:** *See* RIBAVIRIN.

**ribose:** Pentose sugar present in RNA (Bryant et al., 1999).

**ribose nucleic acid:** RNA; often single-polynucleotide chains principally as mRNA, tRNA, and rRNA. Double-stranded RNA is involved in viral replication (Hull, 2002) and gene silencing (Soll et al., 2001; Alberts et al., 2002). *See* RNA INTERFERENCE.

**ribosomal RNA:** rRNA; RNA derived primarily from rRNA genes in the nucleolus and a group of repeat (5S RNA) genes further along the chromosome. The initial product is a single 45S RNA, which is further processed to give 28S, 18S, and 5.8S RNAs. Together with proteins, 18S forms the small ribosomal subunit, and 28S, 5.8S, and 5S RNAs the larger subunit (Alberts et al., 2002).

**ribosome:** Structure composed of rRNA and proteins forming the site of protein synthesis. Ribosomes occur free in the cytosol or attached to rough endoplasmic reticulum during the synthesis of proteins for retention in endoplasmic reticulum membranes or for secretion (Spirin, 2000). Some antibiotics act by inhibiting mRNA translation (Alberts et al., 2002; Walsh, 2003; Nierhaus and Wilson, 2004). *See* ANTIBIOTICS.

**ribulose bisphosphate:** RuBP; carbon dioxide accepter in the first step of the Calvin cycle for photosynthesis (Taiz and Zeifler, 2002). *See* PHOTOSYNTHESIS.

**ribulose bisphosphate carboxylase:** Syns. Rubisco, carboxy-dismutase; a molecule that catalyzes the addition of either carbon dioxide or oxygen to ribulose bisphosphate, a key step in photorespiration. It forms a major protein in chloroplasts (Raghavendra, 1998; Rao, 1999). Rubisco activity is promoted by low sucrose and inhibited by high sucrose in plants in vitro (Dalton, 1980; Roh and Choi, 2004). *See* IN VITRO WEANING; MICROPLANT ESTABLISHMENT; PHOTOSYNTHESIS.

**rifampicin:** Antibiotic inhibiting prokaryotic RNA polymerases and their association with DNA (Walsh, 2003). *See* ANTIBIOTICS.

**RITA**: A patent ebb-and-flow culture vessel system for plant tissues. *See* TEMPORARY IMMERSION.

**RNA:** *See* RIBOSE NUCLEIC ACID.

**RNAi:** *See* RNA INTERFERENCE.

**RNA interference:** RNAi; double-stranded interference RNA involved with gene silencing through the binding of one stand of the RNAi either directly to the DNA or to mRNA, which is subsequently destroyed (Matzke and Matzke, 2000; Hannon, 2003; Novina and Sharp, 2004). RNAi is involved also with the formation of heterochromatin (Martienssen et al., 2004). *See* GENE SILENCING; HETEROCHROMATIN.

**robotics:** Use of computer-controlled machines in production. Prototype robotic arms have been developed to cut and place explants during automated micropropagation (Fujita and Kinase, 1991). *See* AUTOMATION.

**Roentgen:** Syn. Roentgen ray; X-ray.

**rogue:** Variant plant in a population; undesirable phenotypic variant; critical evaluation and elimination of unwanted plants from a population.

*rol* **genes:** Genes on the Ri plasmid of *Agrobacterium rhizogenes.* Transformation with these genes confers increased sensitivity to auxin, leading to root formation (Slater et al., 2003). *See AGROBACTERIUM RHIZOGENES;* ROOTING.

**root:** Lower, descending part of the main plant axis; usually below ground to anchor plant and obtain water and dissolved minerals (Esau, 1977; Anderson et al., 1997). *See* ROOT CULTURE.

**root culture:** Isolated main or lateral root tips in vitro. Whereas some root cultures show indeterminate growth habits, many stop growing after one or a few subcultures (George, 1993).

**root hair:** Epidermal cell outgrowths from root maturation zone aiding the uptake of water and dissolved minerals. Root hairs do not usually occur on roots developed in water or on culture medium (Esau, 1977).

**rooting:** Process of initiation and development of roots (Anderson et al., 1997).

**root nodules:** Nodules present on the roots of a small number of plants, especially legumes, where nitrogen-fixing bacteria of the genus *Rhizobium* are symbiotic with the host (Mauseth, 1998). *See* NITROGEN FIXATION.

**rootstock:** Upright underground stem of plant to which scion is grafted (Hartmann et al., 2001). *See* MICROGRAFTING.

**rotary shaker:** Variable-speed platform shaker with a circular motion for agitating culture flasks. *See* LIQUID CULTURE.

**Roundup:** Trade name for the herbicide glyphosate.

**rRNA:** *See* RIBOSOMAL RNA.

**Rubisco:** *See* RIBULOSE BISPHOSPHATE CARBOXYLASE.

**RuBP:** *See* RIBULOSE BISPHOSPHATE.

**RuBPase:** *See* RIBULOSE BISPHOSPHATE CARBOXYLASE.

**rudiment:** Initial developmental stage of an organ (primordium) or arrested development of an organ (vestige).

**runner:** A horizontal branch rooting at the nodes and behind the shoot tip to yield daughter plants by vegetative propagation, e.g., in strawberry (Hartmann et al., 2001).

**ruthenium red:** Temporary stain for detecting pectins (Gahan, 1984).

**salicyclic acid:** SA; associated with biotic stress signaling in plants (Strange, 2003). *See* INDUCED RESISTANCE; SYSTEMIC ACQUIRED RESISTANCE.

**sand:** Loose material consisting of small rock and mineral particles, distinguishable by the naked eye, with a diameter range from 0.062 to 2 mm.

**saprophyte:** Organisms (usually fungus) living on dead and decaying tissue (Tate, 2000; Watkinson and Gooday, 2001).

**satellite DNA:** *See* MICROSATELLITES.

**scanning electron microscopy:** *See* ELECTRON MICROSCOPE.

**scion:** A cutting from the upper portion of a plant that is grafted onto the rootstock (stock) of another, usually closely related, species (Hartmann et al., 2001). *See* MICROGRAFTING; REJUVENATION; ROOTSTOCK.

**screening:** *See* CULTIVABLE CONTAMINANTS.

**scRNA:** Small cytosolic RNA; associated with proteins as part of the signal recognition particle during protein synthesis on the rough endoplasmic reticulum (Soll et al., 2001; Alberts et al., 2002).

**secondary embryos:** Adventitious embryos that develop on previously formed embryos (Bajaj, 1995).

**secondary meristem:** *See* CAMBIUM.

**secondary metabolites:** Plant compounds that are not directly involved in plant growth and development but are involved in protection against pests and diseases and suppressing competitors (Bonnett and Glasby, 1991; Dey and Harborne, 1996). *See* ALLELOPATHY; CELL SUSPENSION CULTURE.

**secondary products:** *See* SECONDARY METABOLITES.

**sectorial chimeras:** *See* CHIMERA.

**seed:** Consists of the bipolar embryo, may contain a nutritive endosperm, and has a protective cell wall. The seed develops from the ovule after fertilization (Esau, 1977). *See* ARTIFICIAL SEED.

**seed banks:** *See* GENE BANKS.

**seed coat:** The testa, a protective coat around the seed (Esau, 1977). *See* ARTIFICIAL SEED.

**seed tubers:** Tubers of a size and grade specified by certification authorities (Cassells, 1997). *See* MICROTUBERS; MINITUBERS.

**segregation:** The separation of homologous chromosomes at anaphase 1 of meiosis resulting in the separation of parental alleles in gamete formation (Grant, 1975).

**selectable marker:** A gene that confers on the transformed plant resistance to a selective agent in the medium, often kanamycin, thereby preventing the growth of nontransformed shoots (Slater et al., 2003). *See* GENETIC ENGINEERING.

**SEM:** *See* ELECTRON MICROSCOPE.

**semiochemicals:** Chemicals that mediate interactions between organisms (Petroski et al., 2004). *See* VOLATILES.

**semisolid medium:** Medium containing reduced amount of gelling agent, so that it is soft as opposed to rigid, allowing good contact between the cells or tissue and the nutrients. *See* PROTOPLAST ISOLATION.

**sequestrene:** A commercial iron chelate generally taken up via the roots of plants in vivo and also used in plant tissue culture media. *See* CHELATING AGENTS; PLANT TISSUE CULTURE MEDIA.

**serologically specific electron microscopy:** *See* IMMUNOSORBENT ELECTRON MICROSCOPY.

**serological tests:** Methods involving the use of antibodies directly (diffusion tests) or linked to latex, immunogold particles, or enzymes to enhance detection of the antigen (Hari and Das, 1998). *See* ANTIGEN; ANTIBODY; ENZYME-LINKED IMMUNOSORBENT ASSAY; IMMUNOSORBENT ELECTRON MICROSCOPY; LATEX AGGLUTINATION.

**sexual reproduction:** The formation of a new individual by the combination of gametes from two parents. Sexual reproduction is a mechanism for recombining genes within the gene pool (Russell, 2002). *See* ASEXUAL REPRODUCTION; MEIOSIS; SEGREGATION.

**shade plants:** Understory plants in a forest adapted to grow at low light levels that may wilt and die if suddenly exposed to bright light (Hartmann et al., 2001). *See* LIGHT PLANTS.

**shading:** Reduction of the light in a greenhouse, commonly by use of blinds or by painting the structure with opaque material (Hartmann et al., 2001). *See* MICROPLANT ESTABLISHMENT.

**shake cultures:** Cultures aerated by agitation of the liquid medium (George, 1993). *See* BIOREACTOR.

**shaker:** A moving platform to aerate cultures. *See* ROTARY SHAKER; SHAKE CULTURES

**shear stress:** The force exerted on cells and microorganisms in culture by the stirring mechanism (Scragg, 1991a). *See* BIOREACTOR.

**shoot apex:** *See* APICAL MERISTEM.

**shoot cultures:** Cultures that lack roots. *See* MICROPROPAGATION.

**shoot tip culture:** *See* MERISTEM CULTURE.

**short-day plants:** Plants that flower only during short days or in which flowering is accelerated by short days (Taiz and Zeigler, 2002). *See* LONG-DAY PLANT.

**silicon:** Si; element that is not essential for plant growth but accumulates in the walls of many grasses (Marschner, 1994).

**silicon-carbide-fiber-mediated transformation:** A method for transforming cells and embryos involving the vortexing of the fibers with the transgene with the plant cells; the fibers enter the cells through wounds (Slater et al., 2003; Li and Gray, 2005). *See* GENE TRANSFER METHODS.

**silver:** Ag; an inhibitor of ethylene (Taiz and Zeigler, 2002). *See* ethylene inhibitors.

**silver nitrate:** An inhibitor of ethylene action incorporated into tissue culture media (George, 1993). *See* ETHYLENE INHIBITORS.

**silver thiosulfate:** An inhibitor of ethylene action incorporated into plant tissue culture media (George, 1993). *See* ETHYLENE INHIBITORS.

**slow growth:** *See* CRYOPRESERVATION; GERMPLASM CONSERVATION.

**snoRNA:** Small nucleolar RNA; involved with RNA processing (Soll et al., 2001; Alberts et al., 2002).

**snRNA:** Structural and regulatory RNA involved with splicing (Soll et al., 2001; Turner, 2001; Alberts et al., 2002).

**sodium:** Na; low-concentration cation in the cytosol that can freely enter the cell through the plasmalemma down an electrochemical gradient. Large amounts of adenosine triphosphate (ATP) are expended pumping $Na^+$ out of the cell and at the same time bringing in $K^+$ against its chemical gradient via a $Na^+/K^+$ ATPase pump (Marschner, 1994; Alberts et al., 2002).

**sodium alginate:** *See* ALGINATE.

**solidifying agents:** *See* GELLING AGENT.

**solid medium:** A medium that has been set with a gelling agent. *See* LIQUID MEDIUM; SEMISOLID MEDIUM; TISSUE SUPPORTS.

**solvents:** Liquids in which chemicals dissolve. In tissue culture, some plant growth regulators are of low solubility in water and must

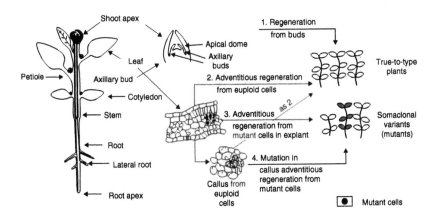

FIGURE 27. The origins of somaclonal variation from adventitious shoots arising from sectors of aberrant cells in the explant. Somaclonal variation may also oc-cur in adventitious shoots arising from mutant cells in callus induced by wound stress or stress induced by the medium or culture environment (Cassells and Curry, 2001). Shoots arising directly from apical meristems and lateral buds are genetically stable except where the plant is a chimera. *See* CHIMERA, ENDO-POLYPLOIDY, SOMACLONAL VARIATION.

be dissolved in other solvents, such as dimethyl sulfoxide, that are miscible with water. Caution: controls should be used to determine any solvent effects. *See* DIMETHYL SULFOXIDE.

**somaclonal variation:** Variation occurring at high frequency in plants arising from tissue culture (Figure 27). Although virus elimi-nation and the breakdown of chimeras may lead to variation in plants derived from tissue culture (Cassells, 1992), the term somaclonal variation is generally used to describe mutation and epigenetic varia-tion (Mohan Jain et al., 1998). Somaclonal variation is associated with adventitious regeneration and depends on the genotype, ontogeny of the explants, and in vitro media and environmental factors. Somaclonal mutants resemble induced mutants (Cassells and Curry, 2001; Joyce et al., 2003). Somaclonal variation has been exploited to produce improved genotypes (Bajaj, 1990, 1996b; Cassells, 1998; Mohan Jain et al., 1998; Jayasankar and Gray, 2005). *See* EPIGENETIC VARIATION; MICROPROPAGATION; MUTATION; MUTATION BREEDING; SPONTANEOUS MUTATION.

**somaclone:** A clone arising from somaclonal variation. *See* SOMACLONAL VARIATION.

**somatic embryo:** An adventitious embryo (Bajaj, 1995). *See* ADVENTITIOUS REGENERATION; ARTIFICIAL SEED.

**somatic embryogenesis:** The development of embryos from a cell or cell aggregate (Bajaj, 1995; Cassells et al., 1997; Bhojwani and Soh, 2001; Thorpe, 2002; Figure 28). *See* ZYGOTIC EMBRYO.

**somatic hybridization:** The fusion of protoplasts, usually to overcome compatibility barriers to sexual hybridization. Also used to transfer or hybridize organellar genomes (Taji et al., 2001; Veilleux et al., 2005). *See* PROTOPLAST FUSION.

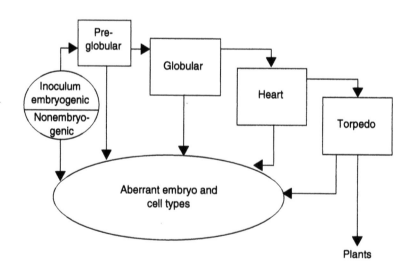

FIGURE 28. Stages in somatic embryogenesis (from Cazzulino et al., 1991). *See* SOMATIC EMBRYOGENESIS.

**Sorbarods:** Commercial cellulose rods for the support of plant tissues in vitro based on the cigarette filter. *See* TISSUE SUPPORTS.

**spontaneous mutation:** Mutation arising in nature or in tissue culture. The frequency of mutation in a character per generation in vivo is of the order of $10^{-4}$ to $10^{-6}$ in eukaryotes (Russell, 2002), whereas when somaclonal variation occurs the mutation rate can be ~30 percent (Karp, 1990). *See* SOMACLONAL VARIATION.

**sporophytic incompatibility:** *See* POSTZYGOTIC INCOMPATIBILITY.

**stable cloning:** The avoidance of somaclonal variation by the selection of an explant, media, and culture environment that will suppress mutation. Cloning based on axillary or nodal bud proliferation is generally recommended; cloning based on adventitious regeneration carries a high risk of somaclonal variation in some genotypes (George, 1993; Cassells and Curry, 2001). *See* MICROPROPAGATION; SOMACLONAL VARIATION.

**stages of micropropagation:** *See* MICROPROPAGATION.

**starch:** A polymer of glucose. Starch is the most abundant storage carbohydrate in plants (Bryant et al., 1999). *See* AMYLOPECTIN; AMYLOSE.

**static liquid medium:** Unshaken liquid medium. *See* BIOREACTOR; SHAKE CULTURES.

**stationary phase:** The phase at the end of the geometric phase of growth of microorganisms or plant cells associated with depletion of the medium. This is the growth stage at which some plant secondary metabolites are produced (George, 1993). *See* BIOREACTOR; GROWTH STAGES.

**stele:** The vascular tissues of a root or shoot but not a leaf (Esau, 1977).

**sterile filtration:** Removal of bacterial and fungal spores from a solution by passing the dissolved compound through a filter, usually of pore size 0.2 μm. Used in the preparation of heat-labile compounds for addition to culture media (George, 1993; Beyl, 2005; Figure 29). *See* MEDIA PREPARATION.

**sterilize:** To make incapable of carrying infection. *See* AUTOCLAVE; INSTRUMENT STERILIZATION; SURFACE STERILIZATION.

**sterilizing agents:** Materials used to eliminate infectious organisms. These include steam at high temperature and pressure for sterilizing instruments, equipment, and media; gamma rays for sterilizing instruments and culture vessels; and chemicals, including hypochlorite, for surface-sterilizing plant tissue. *See* AUTOCLAVE; SURFACE STERILANTS.

**stirred bioreactors:** Bioreactors in which the contents are agitated by a rotating impeller. This serves the two purposes of maintaining the even dispersal of the cells or microorganisms in the medium and of aeration (Scragg, 1991a; Hvoslef-Eide and Preil, 2004). *See* AIRLIFT BIOREACTOR.

**stirred cultures:** Cultures agitated by use of a rotation paddle or impeller (Scragg, 1991a). *See* BIOREACTOR.

**stock plant:** Plant selected as a source of propagative material for either vegetative propagation (Hartmann et al., 2001) or tissue culture (Cassells and Doyle, 2004); also referred to as a mother plant. *See* MICROPROPAGATION; DONOR PLANT.

**stock plant treatments:** *See* MICROPROPAGATION.

**stock solutions:** Concentrated solutions from which dilutions and mixtures are prepared. It is conventional to prepare stock solutions of media components and to mix and dilute these before sterilization and dispensing (George, 1993; Beyl, 2005). *See* MEDIA PREPARATION.

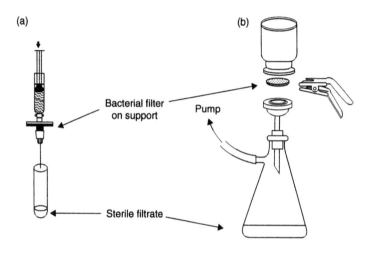

FIGURE 29. Illustration of sterile filtration (a) small scale sterilization; (b) large scale sterilization. This procedure is used to sterilize heat sensitive media components. *See* STERILE FILTRATION.

**stolon:** A modified stem (Esau, 1977) involved in vegetative propagation (Hartmann et al., 2001). The stolon grows horizontally, and when it reaches a suitable habitat, leaf development and rooting may occur, as in strawberry runners. The apical bud of the stolon may swell to form a storage organ, e.g., the potato tuber. *See* STORAGE ORGANS.

**stoma, pl. stomata:** The leaf or stomatal pore including the pair of guard cells (Esau, 1977; Taiz and Zeigler, 2002). Stomatal malfunction resulting in desiccation of microplants at establishment in the greenhouse is associated with high humidity in the culture vessel, which may suppress the calcium uptake in the leaves necessary for functionality (Cassells and Roche, 1994; Cassells and Walsh, 1994).

**storage organs:** Leaves, petioles, shoots, and roots can be modified into storage organs (Esau, 1977; Graham et al., 2003). *See* BULB; CORM; RHIZOME; TUBER.

**strain:** The negative influence of a biotic or abiotic factor on the physiology of a plant that is maintained by homeostatic mechanisms (Lerner, 1999). *See* ABIOTIC STRESS; BIOTIC STRESS.

**stress:** The negative influence of a biotic or abiotic factor on the survival, growth, yield, or reproduction of a plant (Lerner, 1999). *See* ABIOTIC STRESS; BIOTIC STRESS; STRAIN; STRESS CROSS-TOLERANCE.

**stress cross-tolerance:** The phenomenon in which induction of a response to one stress may confer protection against another stress; e.g., exposure to ultraviolet light may protect against pathogen stress (Taiz and Zeigler, 2002; Strange, 2003).

**stress-induced proteins:** Nonconstitutive proteins that are expressed, or constitutive proteins that are up-regulated, following an abiotic or biotic stress (Taiz and Zeigler, 2002). *See* ABIOTIC STRESS; BIOTIC STRESS.

**subculture:** To divide up a culture with the intention of maintaining the material in an active vegetative state or of multiplying the material.

**suberin:** A fatty acid polyester located in the walls of the endodermis and in bark (Brett and Waldron, 1996; Carpita et al., 2001). *See* PLANT CELL WALL.

**sucrose:** A disaccharide of glucose and fructose. Sucrose is the main form in which sugars are transported in plants and in which carbohydrate is stored in sugar cane and sugar beet. Sucrose is the most widely used energy source in heterotrophic plant tissue culture (Bryant et al., 1999). *See* AUTOTROPHIC CULTURE; HETEROTROPHIC GROWTH; MIXOTROPHIC CULTURES.

**sugar:** Usually refers to monosaccharides and sucrose, low-molecular-weight carbohydrates (Bryant et al., 1999).

**sulfur:** S; essential mineral, a component of some amino acids and secondary metabolites (Marschner, 1994). A macronutrient included in media in the form of sulfates (George, 1993). *See* MACRONUTRIENTS; MINERAL NUTRITION.

**supplemented medium:** A standard (published) medium to which additional compounds or increased amounts of existing compounds have been added (George, 1993, 1996).

**supports for tissues:** Alternatives to gelling agents to prevent plant tissues becoming anoxic as a result of submersion in a liquid culture medium. Such supports include cellulose fibers, polyurethane foam, paper bridges, and floating rafts. *See* AERATION; GELLING AGENT; POLYURETHANE FOAM; RAFT; SORBARODS.

**surface sterilants:** Chemicals used to sterilize the surface of explants or ultraviolet light used to sterilize laminar flow and other inoculation cabinets. *See* SURFACE STERILIZATION.

**surface sterilization:** The elimination of microorganisms from surfaces. In tissue culture the process is concerned with both pathogenic and environmental microorganisms, which may be transferred with the tissues into culture. Failure of surface sterilization may be due to the inactivation of the active principle, commonly hypochlorite; the absence of a wetting agent in the sterilizing solution, resulting in poor wetting of the hydrophobic plant surface; the presence of the microorganisms in resistant biofilms; or the partial penetration though natural openings of the tissues by the contaminating microorganisms and the endophytic presence of the contaminating organisms (Cassells and Doyle, 2004). *See* BIOFILMS; ENDOPHYTIC ORGANISMS; GOOD LABORATORY PRACTICE; HEMIENDOPHYTE.

**suspension cultures:** *See* CELL SUSPENSION CULTURE.

**suspensor cells:** The cells that link the embryo to the parental tissue and conduct nutrients to the embryo (Esau, 1977).

**symplastic movement:** Movement via the cytoplasm of the cell and via the plasmodesmata to neighboring cells. *See* APOPLASTIC MOVEMENT; PHLOEM-RESTRICTED MICROORGANISMS.

**synseed:** *See* ARTIFICIAL SEED.

**synthetic seed:** *See* ARTIFICIAL SEED.

**systemic:** A compound or microorganism that is distributed throughout the plant. The term is also used in reference to disease but with the caution that endophytic microorganisms may be spread through the system but show localized accumulation; e.g., they may be phloem restricted (Bove and Garnier, 1997; Hull, 2002). *See* ENDOPHYTIC ORGANISMS; PATHOGEN ELIMINATION.

**systemic acquired resistance:** SAR; resistance mechanism to challenge inoculation, induced by pathogens that cause hypersensitive necrosis involving salicylic acid signaling and accumulation of pathogenesis-related proteins (van Loon, 2000). *See* INDUCED SYSTEMIC RESISTANCE; PATHOGENESIS-RELATED PROTEINS.

**2,4,5-t:** *See* TRICHLOROPHENOXYACETIC ACID.

**T**

**tagged:** Labeled with either an isotope or non-radioactive marker such as a fluorochrome. *See* RADIOLABELING.

**tangential:** Describes a longitudinal section through a cylindrical structure that does not pass through the diameter.

**tannin:** A compound such as polyphenols; able to precipitate gelatin in animal hides as an insoluble complex (Hemingway and Karchesy, 1989). Hydrolyzable tannins are formed from esters and other derivatives of gallic acid which accumulate in plant vacuoles and are deposited in the bark along with condensed tannins which are polymeric flavenoid compounds (Metzler, 2003).

**tapetum:** Layer of cells surrounding the spore mother cells in an anther and a source of food (Esau, 1977).

**taproot:** A persistent primary root often penetrating to some depth below ground. The taproot may act as a storage organ (Esau, 1977).

**tare:** Weight of a weighing container or paper, which is deducted from the overall weight of the object plus the container or paper.

**target cells:** The cells that respond to a plant growth regulator, e.g., the basal cells of young glandular trichomes, which give rise to adventitious shoots in *Kohleria* spp. The concept has been extended to those cells transformed by *Agrobacterium tumefaciens* (Geier and Sangwan, 1996). *See* GENETIC ENGINEERING.

**taxis:** The free movement of whole organisms in response to external stimuli.

**taxon:** Named taxonomic group of any rank (Heywood, 1993; Quicke, 1993).

doi:10.1300/5648_20 *217*

**taxonomic characters:** Features assessed in isolation from the rest of the plant, e.g., form, physiology, structure. *See* PLANT BREEDERS' RIGHTS.

**TCA cycle:** *See* CITRIC ACID CYCLE.

**T-DNA:** That part of the Ti plasmid that is incorporated into the host plant (Glick and Pasternak, 2003). *See* AGROBACTERIUM TUMEFACIENS.

**TDZ:** *See* THIDIAZURON.

**Teepol:** A detergent.

**telocentric:** Chromosome with the centromere close to one end (Dyer, 1979).

**telomere:** The end of the chromosome distal to the centromere. The telomere is associated with heterochromatin and permits completion of the terminal DNA replication during S phase. It prevents chromosomes from joining at the ends (Dyer, 1979; Kipling, 1996). *See* TELOMERE AGING.

**telomere aging:** Experimental evidence indicates that a correlation exists between telomere length and replicative lifespan in human cells in vitro. It is unclear whether critical length, average length, or the opening of the loop structure triggers the signaling cascade that ultimately leads to cell cycle arrest after a fixed number of cell population doublings (Kipling, 1996).

**telophase:** Stage in mitotic and meiotic division following anaphase in which the nuclei reform prior to cytokinesis (Dyer, 1979; Russell, 2002).

**TEM:** Transmission electron microscopy (Murphy, 2001). *See* ELECTRON MICROSCOPE.

**temporary immersion:** Employs the ebb and flow principle used in plant irrigation (Hartmnann et al., 2001) to provide nutrients for tissues in vitro as an alternative to growth on solid media, shaker culture, or in liquid with forced aeration (Alvard et al., 1993; Hvoslef-Eide and Preil, 2004). *See* BIOREACTOR; TISSUE SUPPORTS.

**tendril:** Structure that supports plants by coiling around adjacent objects. It derives from a leaf, leaflet, branch, or inflorescence (Esau, 1977).

**tent:** Plastic or glass cover for plants to maintain a high humidity (George, 1993).

**teratogenic:** Agent inducing teratomas (gross abnormalities) in an individual.

**terminal:** Sited at the apex or tip; the end of life.

**terpene:** *See* TERPENOID.

**terpenoid:** Unsaturated hydrocarbon isoprene units of plant secondary compounds, resins, oils, vitamins A, E, K, and carotenoids (Harborne and Tomas-Barberan, 1991).

**Terramycin:** *See* OXYTETRACYCLINE.

**testa:** Protective seed covering derived from the integuments of the fertilized ovule (Esau, 1977). *See* ARTIFICIAL SEED.

**test cross:** Cross between a homozygous recessive and an individual of uncertain genetic makeup (Allard, 1999).

**test tube:** Sealed glass tube with a single opening, used in tissue culture.

**test-tube fertilization:** In vitro pollination and fertilization (Rangaswamy, 1977).

**tetracycline:** A broad-spectrum antibiotic prepared from the cultures of certain *Streptomyces* spp. (Walsh, 2003). *See* ANTIBIOTICS.

**tetrad:** Set of four cells derived by meiosis (Russell, 2002).

**tetrad analysis:** Genetic analysis of the tetrad to determine the nature and extent of recombination (Russell, 2002).

**tetramethylrhodamine isothiocyanate:** *See* RHODAMINE ISOTHIO-CYANATE.

**tetraploid:** Chromosome number equal to four haploid sets of chromosomes (Allard, 1999).

**tetrasomic:** *See* ANEUPLOID.

**tetrazolium:** A series of compounds that on reduction produce colored formazan products. They are used in quantitative cytochemical studies on dehydrogenases. Triphenyl tetrazolium chloride is reduced to a red formazan by viable seeds but not by nonviable seeds (Gahan, 1984).

**Thallophyta:** Former division containing bacteria, fungi, bryophtes, and pteridophytes.

**thermal imaging:** Deriving images from analysis of temperature (Holst, 2000).

**thermoperiodic response:** Plant's response to diurnal temperature fluctuations.

**thermotherapy:** The use of a heat treatment to inactivate viruses in plants. The treatment can be applied to plants in vivo or in vitro (Mink et al., 1998; Hull, 2002). *See* VIROID ELIMINATION; VIRUS ELIMINATION.

**thiamine:** Vitamin B1; active form thiamine pyrophosphate, a coenzyme involved in carbohydrate metabolism (Metzler, 2003). Thiamine is synthesized by plants but added to tissue culture media as thiamine hydrochloride (George, 1993, 1996). *See* VITAMIN B COMPLEX.

**thidiazuron:** TDZ; a synthetic diphenylurea-type cytokinin developed as a defoliant and herbicide. It is increasingly used in plant tissue culture media for organogenesis and embryogenesis (Murthy et al., 1998). *See* PLANT GROWTH REGULATOR.

**thin cell layers:** Strips of plant tissue one to two cells thick, e.g., epidermal peels used in vitro for adventitious regeneration, often after transformation or for studies on flowering in vitro (Tran Thanh Van, 1973; Tran Thanh Van et al., 1974). *See* EPIDERMIS.

**thin layer chromatography:** A method employing a glass plate or film covered with either alumina or silica gel or cellulose for the separation of components of a mixture placed on it by running a solvent across the surface (Fried and Sherma, 1999; Sherma and Fried, 2003).

**thiolutin:** An antibiotic inhibitor of the elongation of mRNA chains in transcription. *See* ANTIBIOTICS.

**2-thiouracil:** Antiviral agent occasionally included in tissue culture media (Cassells, 1983; Hull, 2002). *See* VIRUS ELIMINATION.

**thiourea:** A reducing agent used to stimulate germination in seeds with a light requirement. It is added to tissue culture media as a reduced nitrogen source (George, 1993, 1996). *See* PLANT TISSUE CULTURE MEDIA.

**thylakoid:** Photosynthetic pigment-bearing vesicular component of grana in chloroplasts (Raghavendra, 1998; Rao, 1999). *See* PHOTO-SYNTHESIS.

**thymidine:** Nucleotide frequently used as $H^3$-thymidine to specifically identify sites of DNA synthesis and to time the phases of the cell cycle. Two millimolar thymidine blocks the cell cycle at the G1/S boundary (Russell, 2002).

**thymine:** Pyrimidine nitrogen base present in DNA but not in RNA (Alberts et al., 2002).

**Thysanoptera:** Thrips; an order of small, sap-sucking insects (Lewis, 1998).

**TIBA:** *See* TRIIODOBENZOIC ACID.

**tin:** Sn; nonessential element (Marschner, 1994).

**Ti plasmid:** *See AGROBACTERIUM TUMEFACIENS.*

**tissue:** Organized groups of cells, e.g., leaf tissue.

**tissue culture:** The culture of tissue pieces, sometimes referred to as complex explants. The origin of the explant within the tissue, the age of the tissue, and the developmental stage of the plant may all influence the tissue response in vitro. It is also important to recognize that not all of the cells in the explant may have the same competence (Stafford and Warren, 1991; George, 1993, 1996; Trigiano and Gray, 2000, 2005; Chawla, 2002; Razdan, 2002). *See* ADVENTITIOUS REGENERATION; SOMACLONAL VARIATION; TARGET CELLS.

**tissue culture contamination:** The presence of cultivable microorganisms and insect pests in tissue cultures. If pathogen-free cultures are established in Stage 1 of micropropagation, then the contaminants will be common environmental microorganisms, mites, and thrips (Cassells, 1997; Cassells et al., 2000; Leifert and Cassells, 2001; Cassells and Doyle, 2004; Figure 30). *See* GOOD LABORATORY PRACTICE; MICROPROPAGATION; PATHOGEN ELIMINATION.

**tissue culture laboratory:** *See* MICROPROPAGATION LABORATORY.

**tissue culture supports:** Alternatives to gelling agents for the prevention of tissue anoxia in liquid culture media. They vary from simple filter paper bridges to floating rafts (George, 1993, 1996) to vessels built for nutrient ebb-and-flow culture. *See* RITA.

**tissue culture vessels:** The containers used for tissue culture. These include test tubes, glass jars, and plastic containers. The lids may be airtight or allow free movement of gases or be of plastics that are differentially permeable to gases (Cassells and Roche, 1994; Cassells and Walsh, 1994). *See* HYPERHYDRICITY; MICROPLANT ESTABLISHMENT; MICROPLANT QUALITY.

**tissue explant:** A piece excised from a tissue.

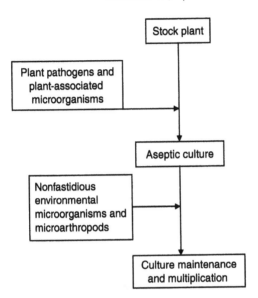

FIGURE 30. Production of pathogen- and contaminant-free cultures depends on establishing pathogen-free plants and following good laboratory practice to manage laboratory contamination (Cassells and Doyle 2004, 2005). *See* Figures 13, 21, 22, Table 2 *and* GOOD LABORATORY PRACTICE; PATHOGEN ELIMINATION; LABORATORY CONTAMINATION MANAGEMENT.

**tissue supports:** Gelling agents and materials that prevent the tissue from becoming anoxic due to submersion in a culture medium. The most frequently used gelling agent is agar because of its relatively low cost and the absence of side effects such as hyperhydricity (Cassells and Collins, 2000). Physical supports include tissue rafts, paper bridges, and polyurethane foam (George, 1993). *See* ANOXIA; FILTER PAPER BRIDGES; POLYURETHANE FOAM; RAFT.

**tmRNA:** *See* TRANSPORT/MESSENGER RNA.

**toluidine blue:** Alkaline aniline stain commonly used in staining semithin sections of resin-embedded material for viewing by light microscopy.

**tonoplast:** The membrane separating the vacuole from the cytoplasm (Leigh and Sanders, 1998).

**topophysis:** An effect of position upon the observed characteristics of a plant, e.g., similar buds from the top and bottom of a plant act differently in vitro.

**torpedo-shaped embryo:** The third stage in embryogenesis (Graham et al., 2003). *See* EMBRYOGENESIS.

**totipotency:** The potential of an isolated undifferentiated plant cell to regenerate into a plant (George, 1993). *See* COMPETENCE.

**toxic:** Poisonous. *See* $LD_{50}$.

**trace element:** *See* MICRONUTRIENTS.

**tracer:** A radiolabeled atom or molecule that can be tracked in a metabolic pathway or in an organism. *See* RADIOLABELING.

**tracheary element:** *See* TRACHEID.

**tracheid:** A cell in the xylem lacking a perforation plate and variable in size (Esau, 1977). *See* XYLEM.

**trait:** *See* CHARACTER.

**transcription:** Copying of a DNA sequence into mRNA to permit information transfer to the cytoplasmic ribosomes for protein synthesis (Alberts et al., 2002). *See* PROTEIN SYNTHESIS.

**transduction:** The transfer of genes between bacteria via a bacteriophage vector (Russell, 2002).

**transfer RNA:** tRNA; RNA carrying amino acids to ribosomes for protein synthesis (Soll et al., 2001; Alberts et al., 2002).

**transformation:** *See* GENETIC ENGINEERING.

**transgene:** A gene introduced by genetic engineering. *See* GENETIC ENGINEERING.

**transition:** *See* MUTATION.

**translation:** The process by which the information in the mRNA is used to produce a polypeptide chain on the ribosome (Alberts et al., 2002). *See* PROTEIN SYNTHESIS.

**translocation:** The movement of water, solutes, and growth regulators within the plant (Ridge, 2002).

**transmission electron microscopy:** TEM; (Murphy, 2001). *See* ELECTRON MICROSCOPE.

**transpiration:** The evaporation of water from a plant surface. Approximately 90 percent of water transpiration from a plant is via the stomata. Transpiration, which is regulated by the stomatal aperture, is important in cooling the plant (Taiz and Zeigler, 2002). *See* MICROPLANT ESTABLISHMENT; STOMA.

**transplanting:** Replanting from a greenhouse to the field or to a larger pot (Hartmann et al., 2001).

**transport/messenger RNA:** tmRNA; RNA involved in releasing ribosomes that are unable to complete protein synthesis on mRNA lacking the stop codon (Zvereva et al., 2000).

**transposable genetic element:** *See* TRANSPOSON.

**transposon:** A DNA sequence that is capable of moving in and out of a chromosome or plasmid (McClintock, 1951). Transposon insertion may result in gene silencing; transposon relocation may result in gene expression (Capy, 1998).

**tree:** A woody perennial plant at least 4 m in height. Micro-propagation of woody plants is a specialization within micro-propagation and may involve problems associated with rejuvenation/reinvigoration and in vitro rooting (Dirr et al., 1987; George, 1993, 1996; Mohan Jain and Ishii, 2003).

**trichlorophenoxyacetic acid:** 2,4,5-T; an auxin-type herbicide banned because of the high level of dioxin present. *See* AUXINS.

**triiodobenzoic acid:** 2,3,5-Triiodobenzoic acid; an auxin transport inhibitor. *See* AUXINS.

**trimethroprim:** An antibiotic inhibitor of folic acid synthesis, used in combination with sulfamethoxazole (Walsh, 2003). *See* ANTI-BIOTICS.

**2,4,5-triphenyltetrazolium chloride:** A colorless vital stain used in protoplast isolation and cryopreservation. It is metabolized in living cells to give a red pigment and is used to test the viability of embryos in seeds (Gahan, 1984). *See* CRYOPRESERVATION; PROTOPLAST ISOLATION.

**tRNA:** *See* TRANSFER RNA.

**true-to-type:** Identical to the parent plant, a clone. *See* SOMACLO-NAL VARIATION.

**tryptone:** Peptides from the controlled enzymatic digestion of casein.

**tuber:** A swollen part of a stem (e.g., potato) or root (e.g., sweet potato). A stem tuber has buds ("eyes") (Esau, 1977).

**tungsten light:** A source of red light to balance the wavelength of the light in growth rooms. *See* ARTIFICIAL LIGHT.

**tunica:** The L1 and L2 layers of the apical meristem (see Figures 7 and 8) *See* APICAL MERISTEM; CORPUS.

**turgor pressure:** The pressure exerted by the protoplast of a cell on the cell wall. It is equal and opposite to wall pressure and it prevents the elastic wall from contracting inward (Ridge, 2002). *See* OSMOTIC POTENTIAL; PROTOPLAST ISOLATION.

**tween:** A wetting agent.

**two-phase medium:** Produced by adding a liquid medium over a semisolid medium to improve the performance of the cultures or to avoid the need to subculture to a rooting medium. A two-phase medium is used to replenish nutrients (Aitken-Christie and Jones, 1985; Maene and Debergh, 1985).

**type I callus:** A type of callus in graminaceous monocots. It gives rise to somatic embryos arrested at the coleptilar or scutellar stage of embryogenesis (Smith, 2000).

**type II callus:** A type of callus in graminaceous monocots. It gives rise to somatic embryos arrested at the globular stage of embryogenesis (Smith, 2000).

**ultrafiltration:** Separation of solutes by passage through a membrane with defined pore size (cutoff) ranging from 100 to 500,000 Da. The solution is pressurized, typically to between 10 and 70 psi. Solutes smaller than the cutoff emerge as ultrafiltrate and retained molecules are concentrated on the pressurized side of the membrane. Compressed nitrogen and peristaltic pump systems are commonly used to pressurize the system (Ghosh, 2003).

**ultraviolet irradiation:** Exposure to ultraviolet light. *See* MUTATION BREEDING; ULTRAVIOLET LIGHT.

**ultraviolet light:** UV light; light of <380 nm, not visible to the human eye. Ultraviolet light is used to sterilize laminar flow cabinets (George, 1993) and to induce mutation (van Harten, 1998). It excites some fluorescent compounds and is used in UV absorbance microscopy to weigh cell components, e.g., nucleic acids and proteins (Gahan, 1984; Murphy, 2001).

**undefined medium:** A medium containing coconut milk or other uncharacterized or only partially characterized components (George, 1993). *See* PLANT TISSUE CULTURE MEDIA.

**undifferentiated cell:** An unspecialized cell such as those found in callus. Such cells may be totipotent (George, 1993). *See* TOTIPOTENCY.

**unorganized tissue:** Tissue in which the cells are undifferentiated. *See* DIFFERENTIATION; ORGANOGENESIS.

**UV light:** *See* ULTRAVIOLET LIGHT.

**vacuole:** A fluid-containing cavity surrounded by a membrane, the tonoplast, and formed initially from the endoplasmic reticulum via small vacuoles, which eventually fuse in fully differentiated cells to form a large central vacuole. Variously contains a variety of molecules including sugars, soluble polysaccharides and proteins, amino acids, anthocyanins, oil droplets, and crystals. The major function of the vacuole is its role in water relations and the maintenance of turgor pressure in cells. The vacuole contains acid hydrolases and acts as a large lysosome, digesting unwanted macromolecules and redundant portions of cytoplasm together with organelles (Leigh and Sanders, 1998; De Deepesh, 2000).

**vacuum:** Void remaining after the evacuation of air from a container. Usually achieved with the aid of a vacuum pump.

**valency:** The number of antigen binding sites per antibody molecule. *See* ANTIBODY.

**valine:** Nonpolar amino acid (Alberts et al., 2002) found in proteins and seeds, often added to plant tissue culture medium (George, 1993, 1996). *See* PLANT TISSUE CULTURE MEDIA.

**vancomycin:** A bacterial cell-wall-inhibiting antibiotic from *Streptomyces orientalis* (renamed *Amycolatopsis orientalis*). The main target is the D-alanyl-D-alanine terminal dipeptide of peptidoglycan precursors; it inhibits the joining of peptidoglycan precursors in bacterial wall formation. Vancomycin binds with the substrate, not the enzyme, in contrast with the mode of action of penicillin (Walsh, 2003). *See* ANTIBIOTICS.

**variant:** *See* MUTANT LINES.

**variation:** Phenotypic differences between individuals of a clone or species resulting from either genetic or environmental effects.

doi:10.1300/5648_22

**variegation:** Patches or streaks of different colored tissues within a plant organ (often leaf or petal). Variegation is caused by either viruses or mineral deficiency or physiological/genetic differences between adjacent cells. *See* CHIMERA.

**variety:** Traditionally, morphologically/phenotypically different individuals grouped together within a species, perhaps occurring in different geographical situations. Now varieties may be distinguished on the basis of subliminal characteristics such as transgenic herbicide resistance (Berman, 2002). *See* PLANT BREEDERS' RIGHTS.

**vascular bundle:** Strands of primary vascular tissue forming the vascular system of the plant, consisting of xylem and phloem, which are often separated by a fascicular cambium (Esau, 1977). *See* CAMBIUM; PHLOEM; XYLEM.

**vascular cambium:** Cambium found in plants exhibiting secondary growth and formed by a lateral meristem that gives rise to the secondary xylem and phloem. It contains fusiform initials, which give rise to both the axial system and the ray initials, which are responsible for the radial aspect of the secondary tissues (Esau, 1977).

**vascular cylinder:** *See* STELE.

**vascular elements:** *See* PHLOEM; XYLEM.

**vascular system:** The continuous water- and solute-conducting tissue of the plant, comprised of the phloem and xylem (Esau, 1977). The vascular system may also carry microorganisms (Hull, 2002; Strange, 2003). *See* PHLOEM; XYLEM.

**vector:** An agent carrying either a pathogen or pollen from one individual to another, e.g., animals, rain splash, contaminated tools, water, wind (Dijkstra and de Jager, 1998; Hull, 2002); also a construct carrying a transgene (Slater et al., 2003). *See* GENETIC ENGINEERING.

**Vegbox container:** A commercial plastic tissue culture container. *See* TISSUE CULTURE VESSELS.

**vegetative growth:** Growth phase of a plant in which preparation for reproduction occurs (Howell, 1998; Hartmann et al., 2001).

**vegetative phase:** The phase in plant growth before the reproductive phase (Howell, 1998). *See* GROWTH STAGES; ONTOGENY.

**vegetative propagation:** Propagation of plants by ways other than sexual reproduction. It uses vegetative propagules and also cuttings and grafting. The term also encompasses in vitro plant propagation methods (Hartmann et al., 2001). *See* MICROPROPAGATION.

**vegetative reproduction:** Asexual reproduction by which specialized multicellular structures are formed by, and separated from, the parent plant, e.g., bulbs, bulbils, corms, gemmae, rhizomes, stems, tubers (MacMillan Browse, 1979; Hartmann et al., 2001).

**vein:** One or a group of vascular bundles in leaves (Esau, 1977). *See* VENATION.

**veination:** *See* VENATION.

**venation:** The pattern formed by veins in a leaf. Dicotyledons tend to have branched veinous systems, whereas monocotyledons generally have parallel veins (Esau, 1977).

**vermiculite:** A variety of minerals, e.g., modified micas, whose granules are very absorbent and can swell. Vermiculite is often mixed with other media for potting-up.

**vernalization:** Low-temperature treatment of either some seeds to induce germination (stratification) or early flowering or of some plants to induce bud break or flowering (Hartmann et al., 2001; Tazi and Zeigler, 2002).

**Versene:** *See* ETHYLENEDIAMINETETRAACETIC ACID.

**vesicle:** Very small fluid-containing cytoplasmic membrane-bound area of variable shape.

**vesicular-arbuscular fungi:** *See* ARBUSCULAR MYCORRHIZAL FUNGI.

**vessel:** Continuous longitudinal tube comprised of vessel elements. Vessels are present in some ferns, the Gnetales (gymnosperms), and most angiosperms (Esau, 1977).

**vessel element:** A tracheary cell in the xylem that has a lignified secondary cell wall and a terminal perforation plate. Vessel elements are linked together longitudinally to form a vessel (Esau, 1977).

**viability testing:** *See* VITAL STAINING.

**viable:** Living; growing.

**victorin:** A specific host toxin produced by *Helminthosporium victoriae* used to screen oat cells for resistance to the pathogen (Strange, 2003).

**vidarabine:** Adenine arabinoside, an antiviral chemical (Cassells, 1983; Hull, 2002).

**vigor:** Healthy growth. *See* MICROPLANT QUALITY.

**viral genome:** The DNA or RNA contained within the virus particle. For information on plant viral genomes, see Hull (2002).

**virazole:** *See* RIBAVIRIN.

**viroid elimination:** Viroids are eliminated by thermotherapy (Mink et al., 1998).

**viroids:** Small pieces of circular naked RNA causing a range of plant infections, e.g., hop stunt, chrysanthemum chlorotic mottle. The detection of viroids was formerly based on extraction of nucleic acids from the infected plant, nuclease digestion of the host nucleic acids, and identification of the nuclease-resistant viroid genome by gel electrophoresis. Now, detection is based mainly on polymerase chain reaction (Singh and Dhar, 1998; Hadidi et al., 2003). *See* GEL ELEC-TROPHORESIS; POLYMERASE CHAIN REACTION.

**virulence:** The ability of a pathogen to cause a disease. There appears to be a gene-for-gene relationship between crop resistance and pathogen virulence for host-specific pathogens (Strange, 2003).

**virus:** Comprised of either DNA or RNA with a protein coat (capsid) and capable of causing infection. For further information on plant viruses, see Hull (2002).

**virus complex:** The presence of more than one virus in a plant. The viruses may not interact or may interact to increase or decrease symptom severity (Hull, 2002). *See* CROSS-PROTECTION.

**virus elimination:** Procedures to eliminate systemic viruses include chemotherapy, thermotherapy, and meristem culture (Faccioli and Marani, 1998; Mink et al., 1998; Jucker, 2001; Hull, 2002). Subsequently, material must repeatedly test negative to ensure that it is virus free (O'Herlihy and Cassells, 2003). *See* PLANT HEALTH CERTIFICATION.

**virus free:** Describes a healthy plant testing negative for a given virus or viruses. However, this does not guarantee that other viruses are not present (Hull, 2002). *See* PLANT HEALTH CERTIFICATION.

**virus tested:** A plant indexed for specified virus(es). For official certification, officially recognized procedures must be used (Cassells and O'Herlihy, 2003). *See* PLANT HEALTH CERTIFICATION.

**vital dye:** A dye that although toxic in large quantities may be relatively unharmful to a cell at very low concentrations. It is used to test the viability of cells either by dye exclusion from living cells, e.g., Evan's blue, or by uptake and reaction with living cells, e.g., fluorescein diacetate, janus green B, neutral red (Gahan, 1984, 1989).

**vital staining:** A technique to determine cell viability (Gahan, 1984; Warren, 1991). *See* PROTOPLAST VIABILITY; VITAL DYE.

**vitamin:** A compound required in trace amounts for the normal functioning of an organism. With the exception of vitamins C, E, and K, vitamins function as coenzymes (Metzler, 2003). Some vitamins are included in plant tissue culture media (George, 1993, 1996). *See* PLANT TISSUE CULTURE MEDIA.

**vitamin A:** Lipid-soluble vitamin absent from plants, though carotene is comprised of two vitamin A molecules.

**vitamin B1:** *See* THIAMINE.

**vitamin B2:** *See* RIBOFLAVIN.

**vitamin B3:** *See* NICOTINIC ACID.

**vitamin B4:** *See* ADENINE.

**vitamin B5:** *See* PANTOTHENIC ACID.

**vitamin B6:** *See* PYRIDOXINE.

**vitamin B12:** A group of Co-containing compounds known as cyanocobalamins. Cyanocobalamin itself is the main B12 vitamin used in food and nutritional supplements. *See* CYANOCOBALAMIN.

**vitamin B complex:** A water-soluble complex of plant-made vitamins routinely added to plant tissue culture media for growth promotion, especially Vitamins B1, B3, and B4.

**vitamin Bx:** *See* PARA-AMINOBENZOIC ACID.

**vitamin C:** *See* ASCORBIC ACID.

**vitamin D2:** Ergocalciferol; lipid soluble.

**vitamin D3:** Cholecalciferol; lipid soluble.

**vitamin E:** *See* α-TOCOPHEROL.

**vitamin G:** Lipid-soluble. *See* RIBOFLAVIN.

**vitamin H:** *See* BIOTIN.

**vitamin K:** Lipid-soluble quinone used by plants in photosynthetic electron transfer (Rao, 1999).

**vitamin PP:** Lipid-soluble. *See* NICOTINIC ACID.

**vitreous shoots:** Shoots with a glassy, transparent, wet, or swollen look followed by necrosis of shoot tip and leaves (Ziv, 1991; Ziv and Ariel, 1994). *See* HYPERHYDRICITY.

**vitrification:** Process by which shoots become vitreous or glassy (now called hyperhydricity) (Debergh et al., 1992); also the immersion stepwise in increasing concentrations of cryoprotectants prior to cryopreservation to prevent the formation of damaging ice crystals (Cassells, 2002). *See* CRYOPRESERVATION; HYPERHYDRICITY.

**vitropathogen:** A microorganism that causes disease in vitro in micropropagation (Cassells and Doyle, 2004). In many cases endophytic pathogens do not cause visible symptoms in microshoots and microplants in vitro. However, environmental contaminants may overrun the nutrient-rich medium, killing the tissues (Long, 1997).

**volatiles:** Compounds existing as gases under ambient environmental conditions, e.g., the plant growth regulator ethylene. Volatiles may accumulate in sealed vessels or in vessels with differentially gas-permeable lids (Cassells et al., 2003). *See* TISSUE CULTURE VESSELS.

**volatility:** Vaporization of liquid at a relatively low temperature and pressure.

**volume:** Space occupied by a mass or bulk and measured in cubic units.

**volumetric flask:** A glass container precisely graduated for a specific volume and used for the preparation of solutions of exact concentrations; should not be used for storage.

**v/v:** Volume/volume; indicator of ratio of volumes of two solutions mixed together.

**wall regeneration:** The regeneration of the cell wall around an isolated protoplast, cybrid, or somatic hybrid. It is a prerequisite for cell division in protoplasts or fused protoplasts (Warren, 1991). *See* CALCOFLUOR; PROTOPLAST ISOLATION.

**waterlogged:** Saturated with water. In vivo this may be associated with flooding or overwatering (Nilson and Orcutt, 1996; Taiz and Zeigler, 2002); some in vitro morphologies resemble the effects of waterlogging (Joyce et al., 2003). *See* HYPERHYDRICITY.

**water potential:** The chemical potential of water in plants. Water moves within plants from regions of high water potential to regions of lower water potential (Ridge, 2002). *See* OSMOTIC POTENTIAL.

**water quality:** An assessment of the freedom of water from microbial contamination, of the level of minerals present, and of the pH. Water is usually graded as potable or nonpotable. Nonpotable water is generally used for industrial process not requiring good-quality water. It is important that water for tissue culture is of high and consistent quality. This is usually achieved by distillation (using a still with nonmetallic heating elements), by deionization, or by a combination of both. Reverse osmosis is also used for water purification (George, 1993; Beyl, 2005). *See* MEDIA PREPARATION.

**water stress:** Stress caused by too much or too little water (Taiz and Zeigler, 2002). *See* DROUGHT; WATERLOGGED.

**wavelength:** The distance between two points in the same phase in consecutive cycles of a wave. *See* ELECTROMAGNETIC SPECTRUM.

**weaning of plants:** Gradual withdrawal of a plant from dependency on a support system (Hartmann et al., 2001). In the case of microplants this is usually withdrawal from dependency on high humidity and hetero- or mixotrophy in the culture vessel (George, 1996;

doi:10.1300/5648_23

Cassells, 2000). *See* IN VITRO WEANING; MICROPLANT ESTABLISH-MENT.

**weedkiller:** *See* HERBICIDE.

**wet tent:** A microenvironment for the establishment of microplants that provides high humidity and shade (George, 1996). *See* MICRO-PLANT ESTABLISHMENT.

**wetting agent:** A chemical that reduces the surface tension of water and enables it to spread on a hydrophobic surface such as the plant surface, helping powders, waxes, and oils to dissolve for spraying. *See* DETERGENT.

**woody plants:** Plants containing lignified secondary xylem in their stems (Esau, 1977). *See* TREE.

**wound cambium:** A cambium that forms protective tissue at the site of injury to a plant (Esau, 1977; Strange, 2003).

**wounding:** Injury to plant tissue associated with an oxidative burst (Inze and van Montague, 2001). The oxidative burst associated with the excision of plant tissue may be an important trigger for cell division in the tissues (Joyce et al., 2003). *See* OXIDATIVE STRESS.

**wound reaction:** *See* WOUND CAMBIUM.

**wound response:** The sequence of events from the oxidative burst when the tissue is damaged to the formation of wound callus or cork, followed in some cases by adventitious root or shoot formation (Esau, 1977; George, 1993; Cassells and Curry, 2001; Inze and van Montague, 2001). *See* ADVENTITIOUS REGENERATION.

**WPM:** A frequently used medium for woody plants. *See* LLOYD AND MCCOWN MEDIUM.

**xanthophyll:** Carotene-derived hydrocarbons functioning as accessory pigments in plants and as a primary light-absorbing pigment in some algae during photosynthesis (Hendry, 1993). *See* PHOTOSYNTHESIS.

**xerograft:** *See* HETEROGRAFT.

**X irradiation:** *See* MUTATION BREEDING.

**X-ray:** Roentgen ray (r); short wavelength ($10^{-1}$ to $10^{1}$ nm on the electromagnetic spectrum) electromagnetic radiation produced from high-speed electrons impacting on a metal surface under vacuum. X-rays are a mutagenic agent. *See* ELECTROMAGNETIC SPECTRUM; MUTATION BREEDING.

**xylan:** Polysaccharide, composed of xylose units, that is present in hemicelluloses (Bryant et al., 1999). *See* PLANT CELL WALL; PROTOPLAST ISOLATION.

**xylem:** The woody portion of the vascular tissue usually in association with, and internal to, phloem (Esau, 1977). Normally comprised of parenchyma, fibers, fiber-tracheids, tracheids, and vessels (trachea) with cell walls reinforced with lignin, though not all cell types may be present in all xylems. Xylem is differentially distributed between roots and stems and between monocotyledonous and dicot plants. Its primary function is to transport water and dissolved minerals from roots to leaves as part of the transpiration stream, though it also acts to support the plant (Ridge, 2002; Taiz and Zeigler, 2002; Tyree and Zimmermann, 2002; Trigiano and Gray, 2005). Primary xylem comprises proto- and metaxylem derived from the primary meristems. Secondary xylem is derived from the cambium. *See* WOODY PLANTS.

**xylem fibers:** Relatively long sclerenchyma cell with simple pits usually formed directly from meristematic cells (Esau, 1977).

doi:10.1300/5648_24

**xylem parenchyma:** Relatively unspecialized cells with slight lignification of the cell walls (Esau, 1977).

**xylem tracheary elements:** Lignified secondary cell wall, often with no living protoplast on maturing. Longer than a vessel cell component and lacking perforation plates, they have various forms of wall thickening, e.g., spiral, annular scalariform, pitted. They are the only cells conducting water up the plant in some angiosperms and in gymnosperms (except Gnetales) (Esau, 1977).

**xylem vessels:** Continuous longitudinal tubes composed of fused cells lacking a living protoplast at maturity, and with perforation plates between these cells. They are present in some ferns, most angiosperms, and the Gnetales (gymnosperms) (Esau, 1977).

**xylogenesis:** Process of the differentiation of cells to form xylem.

**xylose:** An aldopentose sugar found widely in plants. It is a component of the rare disaccharide primeverose (Bryant et al., 1999). *See* XYLAN.

**Y**

**yeast extract:** Yeast autolysates, i.e., yeast cells that are allowed to die and disintegrate. The yeasts' hydrolytic enzymes break the proteins down to peptides.

**yeast identification:** Based on cell and colony morphological characteristics and color. Environmental yeasts can be identified using API kits (Walker, 1998). *See* API KITS.

**yeasts:** Uniform budding fungi that are not taxonomically uniform (Walker, 1998). Yeasts can be indicators of poor air quality in a tissue culture laboratory (Leifert and Cassells, 2001; Cassells and Doyle, 2004). *See* GOOD LABORATORY PRACTICE; TISSUE CULTURE CONTAMINATION.

doi:10.1300/5648_25

**zeatin:** Naturally occurring cytokinin that with auxin stimulates division in mature cells (Moore, 1989; Taiz and Zeigler, 2002). It is often included in plant tissue culture media (George, 1993, 1996) and is soluble in 0.1M hydrochloric acid.

**zeatin riboside:** Naturally occurring cytokinin. It is the main cytokinin in xylem sap (Moore, 1989) and is occasionally included in plant tissue culture media (George, 1993, 1996). *See* CYTOKININS.

**zeaxanthin:** A carotinoid that absorbs in the blue region. It is a component of xanthophylls that protects chloroplasts against excess light (Hendry, 1993; Rao, 1999). *See* PHOTOSYNTHESIS.

**Zephiran:** A commercial disinfecting agent for plant material; contains benzalkonium chloride (George, 1993). *See* SURFACE STERILANTS.

**Ziehl's stain:** *See* CARBOL FUCHSIN.

**zinc:** Zn; microelement essential for chlorophyll, IAA synthesis, and a number of enzymes associated with plant growth (Marschner, 1994). It is added to plant tissue culture media as either zinc chloride or zinc sulfate (George, 1993, 1996). *See* PLANT TISSUE CULTURE MEDIA.

**zygote:** Diploid cell resulting from the fusion of a male and female gamete (fertilized egg). It develops into a zygotic embryo (Graham et al., 2003; Mauseth, 2003).

**zygotic embryo:** An embryo developed from a fertilized egg cell. *See* EMBRYOGENESIS; SOMATIC EMBRYOGENESIS.

doi:10.1300/5648_26

# References

Ackman, R.G. and Metcalfe, L.D. (1976). *Analysis of Fatty Acids and Their Esters by Chromatographic Methods.* Simpsonville, MD: Preston Publishing.

Agrios, G.N. (1997). *Plant Pathology.* New York: Academic Press.

Aitken-Christie, J. and Jones, C. (1985). Towards automation: Radiata pine shoot hedges in vitro. *Plant Cell Tissue Organ Cult. 8:* 185-196.

Alberts, B., Johnson, A., Lewis, J., Raff, M., Roberts, K., and Walter, P. (2002). *Molecular Biology of the Cell* (4th Ed.). New York: Garland Science.

Alexopoulos, C. J., Mims, C. W., and Blackwell, M. (1979). *Plant Pathology* (4th Ed.). New York: Academic Press.

Allard, R.W. (1999). *Principles of Plant Breeding.* New York: Wiley.

Alvard, D., Cote, F., and Teisson, C. (1993). Comparison of methods of liquid medium cultures for banana micropropagation: Effects of temporary immersion of explants. *Plant Cell Tissue Organ Cult. 32:* 55-60.

Anderson, H.M., Clarkson, D.T., Shewry, P.R., Jackson, M.B., and Barlow, P.W. (1997). *Plant Roots: From Cells to Systems.* Dordrecht, The Netherlands: Kluwer Academic Publishers.

Anonymous (1998). *Epigenetics.* Novartis Foundation. Symposium 214, Vol. 214. New York: Wiley.

Bajaj, Y.P.S. (1990). *Somaclonal Variation in Crop Improvement I.* Berlin: Springer-Verlag.

———. (1995). *Somatic Embryogenesis and Synthetic Seed II. Biotechnology in Agriculture and Forestry,* Vol. 31. Berlin: Springer-Verlag.

———. (1996a). *Plant Protoplasts and Genetic Engineering. Biotechnology in Agriculture and Forestry,* Vol. 9. New York: Springer-Verlag.

———. (1996b). *Somaclonal Variation in Crop Improvement II. Biotechnology in Agriculture and Forestry,* Vol. 36. Berlin: Springer-Verlag.

Baldi, P. and Hatfield, G.W. (2002). *DNA Microarrays and Gene Expression: From Experiments to Data Modelling.* Cambridge: Cambridge University Press.

Balkema, G.H. (1972). Diplontic drift in chimeric plants. *Radiat. Bot. 12:* 51-55.

Balows, A. (1992). *Prokaryotes.* New York: Springer-Verlag.

Barnett, H.L. and Hunter, B.B. (1998). *Illustrated Genera of Imperfect Fungi* (4th Ed.). St. Louis, MO: APS Press.

Barrett, C. and Cassells, A.C. (1994). An evaluation of antibiotics for the elimination of *Xanthomonas campestris* pv. *pelargonii* (Brown) from *Pelargonium* × *domesticum* cv. Grand Slam explants *in vitro*. *Plant Cell Tissue Organ Cult. 36:* 169-175.

Basra, A.S. (2000). *Plant Growth Regulators in Agriculture and Horticulture: Their Role and Commercial Uses.* Binghamton, NY: Haworth Press.

Basra, A.S. and Basra, R.K. (1997). *Mechanisms of Environmental Stress Resistance in Plants.* London: Taylor & Francis.

Bates, G.W. (1989). Electrofusion: The technique and its application to somatic hybridisation. In Bajaj, Y.P.S. (Ed.), *Biotechnology in Agriculture and Forestry, Vol. 9. Plant Protoplasts and Genetic Engineering II* (pp. 241-256). Berlin: Springer-Verlag.

Beasley, C.A. (1977). Ovule culture: Fundamental and pragmatic research for the cotton industry. In Reinert, J. and Bajaj, Y.P.S. (Eds.), *Plant Cell, Tissue and Organ Culture* (pp. 160-178). Berlin: Springer-Verlag.

Beers, E.P. and McDowell, J.M. (2001). Regulation and execution of programmed cell death in response to pathogens, stress and developmental cues. *Curr. Opin. Plant Biol. 4:* 561-567.

Bell, S.M., Gatus, B.J., and Pham, J.N. (1999). *Antibiotic Susceptibility Testing by the CDS Method.* A Concise Laboratory Manual 1999. Sydney: Arthur Productions.

Bergman, L. (1960). Growth and division of single cells of higher plants *in vitro*. *J. Gen. Physiol. 43:* 841-851.

Berman, S. (2002). Plant-specific intellectual property rights. In Vainstein, A. (Ed.), *Breeding for Ornamentals: Classical and Molecular Approaches* (pp. 347-379). Dordrecht, The Netherlands: Kluwer Academic Publishers.

Beyl, C.A. (2005). Getting started with tissue culture—media preparation, sterile technique, and laboratory equipment. In Trigiano, R.N. and Gray, D.J. (Eds.), *Plant Development and Biotechnology* (pp. 19-37). Boca Raton, FL: CRC Press.

Bhojwani, S.S. and Soh, W.-Y. (2001). *Current Trends in the Embryology of Angiosperms.* Dordrecht, The Netherlands: Kluwer Academic Publishers.

Binns, A. and Meins, F. (1979). Cold-sensitive expression of cytokinin-habituation by tobacco pith cells in culture. *Planta 145:* 365-369.

Birren, B.W. and Lai, E. (1993). *Pulsed Field Gel Electrophoresis.* Oxford: Elsevier.

Bohmont, B.L. (2002). *Standard Pesticide User's Guide.* London: Pearson Education.

Bonnett, R. and Glasby, J.S. (1991). *Directory of Plants Containing Secondary Metabolites.* London: Taylor & Francis.

Boone, D.B. and Castenholz, R.W. (1999). *Bergey's Manual of Systematic Bacteriology: The Archea, Cyanobacteria, Phototrophs and Deeply Branching Genera,* Vol. 1. London: Lippincott, Williams & Wilkins.

Bove, J.M. and Garnier, M. (1997). Walled and wall-less eubacteria from plants: Sieve-tube-restricted plant pathogens. In Cassells, A.C. (Ed.), *Pathogen and Microbial Contamination Management in Micropropagation* (pp. 45-60) Dordrecht, The Netherlands: Kluwer Academic Publishers.

Bozzola, J.J. and Russell, L.D. (1998). *Electron Microscopy*. Boston, MA: Jones and Bartlett.

Brandon, C. and Tooze, J. (1998). *Introduction to Protein Structure*. London: Taylor & Francis.

Brett, C.T. and Waldron, K. (1996). *Physiology and Biochemistry of Plant Cell Walls*. London: Routledge.

Brettschneider, R. (1998). RFLP analysis. In Karp, A., Isaac, P.G., and Ingram, D.S. (Eds.), *Molecular Tools for Screening Biodiversity* (pp. 83-95). London: Chapman & Hall.

Briggs, W.R. and Liscum, E. (1997). Blue light-activated signal transduction in higher plants. In Aducci, P. (Ed.), *Signal Transduction in Plants*. Basle: Birkhauser Verlag.

Bryant, J.A., Burrell, M.M., and Kruger, N.J. (1999). *Plant Carbohydrate Biochemistry*. Abingdon, UK: BIOS Scientific.

Bryce, J.H. and Hill, S.A. (1993). Energy Production in Plant Cells. In P.J. Lea and R.C. Leegood (Eds.), *Plant Biochemistry and Molecular Biology*. New York: Wiley.

Buchanan, B.B., Jones, R.L., and Gruissem, W. (2002). *Biochemistry and Molecular Biology of Plants*. New York: Wiley.

Buiatti, M. (1997). DNA amplification and tissue cultures. In Reinert, J. and Bajaj, Y.P.S. (Eds.), *Plant Cell, Tissue and Organ Culture* (pp. 358-374). Berlin: Springer-Verlag.

Buncel, E. and Jones, J.R. (1991). *Isotopes in the Physical and Biomedical Sciences: Labelled Compounds*. Amsterdam: Elsevier.

Campbell, R. (1989). *Biological Control of Microbial Plant Pathogens*. Cambridge: Cambridge University Press.

Candress, T., Hammond, R.W., and Hadidi, A. (1998). Detection and identification of plant viruses and viroids using polymerase chain reaction (PCR). In Hadidi, A., Khetarpal, R.K., and Koganezawa, H. (Eds.), *Plant Virus Disease Control* (pp. 399-416). St. Louis, MO: APS Press.

Capellades, M., Lemeur, R., and Debergh, P. (1991). Effects of sucrose on starch accumulation and rate of photosynthesis in *Rosa* cultured *in vitro*. *Plant Cell Tissue Organ Cult. 25:* 21-26.

Caponetti, J.D., Gray, D.N., and Trioiano, R.N. (2005). History of plant tissue culture. In Trigiano, R.N. and Gray, D.J. (Eds.), *Plant Development and Biotechnology* (pp. 9-15). Boca Raton, FL: CRC Press.

Capy, P. (1998). *Evolution and Impact of Transposable Elements*. Dordrecht, The Netherlands: Kluwer Academic Publishers.

Carpita, N.C., Tierney, M., and Campbell, M. (2001). *Plant Cell Walls*. Dordrecht, The Netherlands: Kluwer Academic Publishers.

Cassells, A.C. (1978). Uptake of charged lipid vesicles by isolated tomato protoplasts. *Nature 275:* 760.

————. (1979). The effect of 2,3,5-triiodobenzoic acid on caulogenesis in callus cultures of tomato and pelargonium. *Physiol. Plant 46:* 159-165.

———. (1983). Chemical control of virus diseases of plants. *Prog. Med. Chem. 21:* 119-155.

———. (1987). *In vitro* induction of virus-free potatoes by chemotherapy. In Bajaj, Y.P.S. (Ed.), *Biotechnology in Agriculture and Forestry, Vol. 3. Potatoes* (pp. 40-50). Heidelberg: Springer-Verlag.

Cassells, A.C. (Ed.) (1988). Symposium on bacterial and bacteria-like contaminants of plant tissue cultures. *Acta Hortic. 225*, ISHS, Wageningen.

Cassells, A.C. (1989). Uptake of viruses by plant protoplasts and their use as transforming agents. In Bajaj, Y.P.S. (Ed.), *Biotechnology in Agriculture and Forestry*, Vol. 9 (pp. 388-405). Berlin: Springer-Verlag.

———. (1991). Setting up a commercial micropropagation laboratory. In Bajaj, Y.P.S. (Ed.), *Biotechnology in Agriculture and Forestry, Vol. 17. High Micropropagation* (pp. 17-31). Berlin: Springer-Verlag.

———. (1992). Micropropagation of commercial *Pelargonium* species and hybrids (glasshouse "Geraniums"). In Bajaj, Y.P.S. (Ed.), *Biotechnology in Agriculture and Forestry, Vol. 20. Micropropagation IV* (pp. 286-306). Berlin: Springer-Verlag.

———. (1997). *Pathogen and Microbial Contamination Management in Micropropagation.* Dordrecht, The Netherlands: Kluwer Academic Publishers.

———. (1998). *In vitro*-induced mutations for disease resistance. In Mohan Jain, S., Brar, D.S., and Ahloowalia, B.S. (Eds.), *Somaclonal Variation and Induced Mutations in Crop Improvement* (pp. 367-378). Dordrecht, The Netherlands: Kluwer Academic Publishers.

———. (2000a). Contamination detection and elimination. In Spier, R.E. (Ed.), *Encyclopedia of Plant Cell Biology* (pp. 577-586). Chichester, UK: Wiley.

———. (2000b). Aseptic microhydroponics: A strategy to advance microplant development and improve microplant physiology. *Acta Hortic. 530:* 187-194.

———. (2001). Contamination and its impacts in tissue culture. *Acta Hortic. 560:* 353-359.

———. (2002). Tissue culture for ornamental breeding. In Vainstein, A. (Ed.), *Breeding for Ornamentals—Classical and Molecular Approaches* (pp. 139-153). Dordrecht, The Netherlands: Kluwer Academic Publishers.

———. (2003). Plant tissue culture: Micropropagation. In Thomas, B., Murphy, D.J., and Murray, B.G. (Eds.), *Encyclopedia of Applied Plant Sciences* (pp. 1353-1360). New York: Academic Press.

Cassells, A.C., Austin, S., and Goetz, E.M. (1987). Variation in tubers in single cell derived clones of potato. In Bajaj, Y.P.S. (Ed.), *Biotechnology in Agriculture and Forestry, Vol. 3. Potatoes* (pp. 375-391). Heidelberg: Springer-Verlag.

Cassells, A.C. and Barlass, M. (1978). A method for the isolation of stable mesophyll protoplasts from tomato leaves throughout the year under standard conditions. *Physiol. Plant. 42:* 236-242.

Cassells, A.C. and Collins, I.M. (2000). Characterization and comparison of agars and other gelling agents for plant tissue culture use. *Acta Hortic. 530:* 203-212.

Cassells, A.C., Croke, J.T., and Doyle, B.M. (1997). Evaluation of image analysis, flow cytometry and RAPD analysis for the assessment of somaclonal variation and induced mutation in tissue culture-derived *Pelargonium* plants. *Angew. Bot. 71*, 125-130.

Cassells, A.C. and Curry, R.F. (2001). Oxidative stress and physiological, epigenetic and genetic variability in plant tissue culture: Implications for micropropagators and genetic engineers. *Plant Cell Tissue Organ Cult. 64:* 145-157.

Cassells, A.C., Deadman, M.L., Brown, C.A., and Griffin, E. (1991). Field resistance to late blight (*Phytophthora infestans* (Mont.) De Bary) in potato (*Solanum tuberosum* L.) somaclones associated with instability and pleiotropic effects. *Euphytica 56:* 75-80.

Cassells, A.C. and Doyle, B.M. (2003). Genetic engineering and mutation breeding for tolerance to abiotic and biotic stresses: Science, technology and safety. *Bulg. J. Plant Physiol* (Special Issue): 38-52.

———. (2006). Pathogen and biological contamination management: The road ahead. In Loyola-Vargas, V.M. and Vasquez-Flota, F. (Eds.), *Plant Cell Culture Protocols* (2nd Ed.) (pp. 35–50). Totowa, NJ: Humana Press.

———. (2005). Indexing for plant pathogens. In Trigiano, R.N. and Gray, D.J. (Eds.), *Plant Development and Biotechnology* (pp. 321-332). Boca Raton, FL: CRC Press.

Cassells, A.C., Doyle, B.M., and Curry, R.F. (Eds.) (2000). Methods and markers for quality assurance in micropropagation. *Acta Hortic. 530*. ISHS, Leuven.

Cassells, A.C., Harmey, M.A., Carney, B.F., McCarthy, E., and McHugh, A. (1988). Problems posed by cultivable bacterial endophytes in the establishment of axenic cultures of *Pelargonium* × *domesticum*: The use of *Xanthomonas pelargonii*-specific ELISA, DNA probes and culture indexing in the screening of antibiotic treated and untreated donor plants. *Acta Hortic. 225:* 153-162.

Cassells, A.C. and Jones, P.W. (1995). *The Methodology of Plant Genetic Manipulation: Criteria for Decision Making*. Dordrecht, The Netherlands: Kluwer Academic Publishers.

Cassells, A.C., Joyce, S.M., O'Herlihy, E.A., Perez-Sanz, M.J., and Walsh, C. (2003). Stress and quality in *in vitro* culture. *Acta Hortic. 625:* 153-164.

Cassells, A.C., Kowalski, B., Fitzgerald, D.M., and Murphy, G.A. (1999). The use of image analysis to study developmental variation in micropropagated potato (*Solanum tuberosum* L.). *Potato Res. 42:* 541-548.

Cassells, A.C. and Long, R.D. (1980a). The regeneration of virus-free plants from cucumber mosaic virus- and potato virus Y-infected tobacco explants cultured in the presence of Virazole. *Z. Naturforsch. 35c:* 350-351.

———. (1980b). Tobacco explant culture as a screening system for antiviral chemicals. In Ingram, D.S. and Helgeson, J.P. (Eds.), *Tissue Culture Methods for Plant Pathologists* (pp. 131-135). Oxford: Blackwell Scientific Publications.

———. (1982). The elimination of potato viruses, X, Y, S and M in meristem and explant cultures of potato in the presence of Virazole. *Potato Res. 25:* 165-173.

———. (1983). The effect of 1-β-D-ribofuranosyl-1,2,4-triazole-3-carboxamide on morphogenesis in tobacco petiole culture. *Physiol. Plant. 59:* 664-668.

Cassells, A.C. and Morrish, F.M. (1986). Genetic, ontogenetic and physiological factors and the development of a broad spectrum medium for *Begonia rex* Putz. cultivars. *Sci. Hortic. 31:* 295-302.

Cassells, A.C. and Morrish, F.M. (1987). Variation in adventitious regenerants of *Begonia rex* Putz. "Lucille Closon" as a consequence of cell ontogeny, callus ageing and frequency of callus subculture. *Sci. Hortic. 32:* 135-145.

Cassells, A.C. and O'Herlihy, E.A. (2003). Pathogen elimination and contamination management. In Mohan Jain, S. and Ishii, K. (Eds.), *Micropropagation of Woody Trees and Fruits* (pp. 103-128). Dordrecht, The Netherlands: Kluwer Academic Publishers.

Cassells, A.C. and Roche, T. (1994). The influence of the gas permeability of the vessel lid and growth-room light intensity on the characteristics of *Dianthus* microplants *in vitro* and *ex vitrum*. In Lumsden, P.J., Nicholas, J.R., and Davies, W.J. (Eds.), *Physiology, Growth and Development of Plants in Culture* (pp. 204-214). Dordrecht, The Netherlands: Kluwer Academic Publishers.

Cassells, A.C. and Tahmatsidou, V. (1997). The influence of local plant growth conditions on non-fastidious bacterial contamination of meristem-tips of *Hydrangea* cultured *in vitro*. *Plant Cell Tissue Organ Cult. 47:* 15-26.

Cassells, A.C. and Tamma, L. (1985). Ethylene and ethane release during tobacco protoplast isolation and protoplast survival potential *in vitro*. *Physiol. Plant. 66:* 303-308.

———. (1987). Survival and division in protoplasts from tobacco (*Nicotiana tabacum*) depends on the physiological state of the individual donor plant. *Physiol. Plant. 69:* 317-322.

Cassells, A.C. and Walsh, C. (1994). The influence of the gas permeability of the culture lid on calcium uptake and stomatal function in *Dianthus* microplants. *Plant Cell Tissue Organ Cult. 37:* 171-178.

Cassells, A.C., Walsh, C., and Periappuram, C. (1993). Diplontic selection as a positive factor in determining the fitness of mutants of *Dianthus* "Mystere" derived from x-irradiation of nodes in *in vitro* culture. *Euphytica 70:* 167-174.

Cazzulino, D., Pedersen, H., and Chin, C.-K. (1991). Bioreactors and image analysis for scale-up and plant propagation. *Cell Struct. Somat. Cell Genet Plant. 8:* 147-177

Chaleff, R.S. and Parsons, M.F. (1978). Direct selection *in vitro* for herbicide resistant mutants of *Nicotiana tabacum*. *Proc Natl Acad. Sci. USA 75:* 5104-5107.

Chawla, H.S. (2002). *Introduction to Plant Biotechnology*. Enfield, NH: Science Publishers.

Chrispeels, M.J. and Sadava, D.E. (2003). *Plants, Genes, and Crop Biotechnology*. Boston, MA: Jones and Bartlett Publishers.

Christianson, M.L. and Warnick, D.A. (1985). Temporal requirements for phytohormone balance in the control of organogenesis *in vitro*. *Dev. Biol. 112:* 494-497.

Christou, P. (1995). Strategies for variety-independent genetic transformation of important cereals, legumes and woody species utilizing particle bombardment. *Euphytica 85:* 13-27.

————. (1996). *Particle Bombardment for Genetic Engineering of Plants*. New York: Elsevier.

Ciofi, C., Funk, S.M., Coote, T., Cheesman, D.J., Hammond, R.L., Saccheri, I.J., and Bruford, M.W. (1998). Genotyping with microsatellites. In Karp, A., Isaac, P.G., and Ingram, D.S. (Eds.), *Molecular Tools for Screening Biodiversity* (pp. 195-201). London: Chapman & Hall.

Cobb, A.H. and Kirkwood, R.C. (2000). *Herbicides and Their Mechanisms of Action*. Boca Raton, FL: CRC Press.

Coico, R., Sunshine, G., and Benjamini, E. (2003). *Immunology: A Short Course*. New York: Wiley.

Compton, M.E. (2005). Elements of *in vitro* research. In Trigiano, R.N. and Gray, D.J. (Eds.), *Plant Development and Biotechnology* (pp. 55-72). Boca Raton, FL: CRC Press.

Cordier, C., Lemoine, M.C., Lemanceau, P., Gianinazzi-Pearson, V., and Gianinazzi, S. (2000). The beneficial rhizosphere: A necessary strategy for microplant production. *Acta Hortic. 530:* 259-268.

Croke, J.T. and Cassells, A.C. (1997). Dark induction and genetic stability of somatic embryos of zonal geraniums *(Pelargonium x hortorum* Baily). *Angew. Bot. 71*, 119-124.

Curry, R.F. and Cassells, A.C. (1998). Callus initiation, maintenance and shoot induction in potato: Monitoring of spontaneous genetic variability *in vitro* and *in vivo*. In Hall, R.H. (Ed.), *Methods in Molecular Biology, Vol. 7. Plant Cell Culture Protocol* (pp. 31-42). Totowa, NJ: Humana Press.

Cutter, E.G. (1978). *Plant Anatomy I. Cells and Tissues* (2nd Ed.) London: Edward Arnold.

Dalton, C.C. (1980). Photoautotrophy of spinach cells in continuous culture: Photosynthetic development and sustained photoautotrophic growth. *J. Exp. Bot. 31:* 791-804.

D'Amato, F. (1997). Cytogenetics of differentiation in tissue and cell cultures. In Reinert, J. and Bajaj, Y.P.S. (Eds.), *Plant Cell, Tissue and Organ Culture* (pp. 343-357). Berlin: Springer-Verlag.

Danon, A., Delorme, V., Mailhac, N., and Gallois, P. (2000). Plant programmed cell death: A common way to die. *Plant Physiol. Biochem. 38:* 647-655.

Darlington, C.D., and La Cour, L.F. (1976). *The Handling of Chromosomes* (6th Ed.). Plymouth, UK: George, Allen and Unwin.

Davies, W.J. and Santamaria, J.M. (2000). Physiological markers for microplant shoot and root quality. *Acta Hortic. 530:* 363-376.

Davis, T.D. and Hassig, B.E. (1994). *Biology of Adventitious Root Formation*. Dordrecht, The Netherlands: Kluwer Academic Publishers.

Debergh, P., Aitken-Christie, J., Cohen, D., Grout, B., von Arnold, S., Zimmermann, R.H., and Ziv, M. (1992). Reconsideration of the term "vitrification" as used in micropropagation. *Plant Cell Tissue Organ Cult. 30:* 135-140.

Debergh, P.C., Topoonyanont, N., van Huylenbroeck, J., Moreira da Silva, H., and Oyaert, E. (2000). Preparation of microplants for *ex vitro* establishment. *Acta Hortic. 530:* 269-273.

Debergh, P.C. and Zimmerman, R.H. (1991). *Micropropagation: Technology and Application*. Dordrecht, The Netherlands: Kluwer Academic Publishers.

De Deepesh, N. (2000). *Plant Cell Vacuoles: An Introduction*. Collingwood, Victoria, Australia: CSIRO.

De Klerk, G.-J. (2002). Rooting of microcuttings: Theory and practice. *In Vitro Plant 38:* 415-422.

De Nettancourt, D. (1993). Self- and cross-incompatibility systems. In Hayward, M.D., Bosemark, N.O., and Romagosa, I. (Eds.), *Plant Breeding: Principles and Prospects* (pp. 203-228). London: Chapman & Hall.

Denny, H. (2002). Interactions between seed plants and microbes. In Ridge, I. (Ed.), *Plants* (pp. 275-324). Oxford: Oxford University Press.

Dey, P.M. and Harborne, J.B. (1996). *Plant Biochemistry*. New York: Elsevier Science.

Dijkstra, J. and de Jager, C.P. (1998). *Practical Plant Virology*. Berlin: Springer-Verlag.

Dirr, M.A., Heuser, C.W., and Dirr, B.L. (1987). *Reference Manual of Woody Plant Propagation: From Seed to Tissue Culture*. Omaha, NE: Varsity Press.

Dix, P.J. (1990). *Plant Cell Line Selection*. Weinheim, Germany: VCH.

Dodds, J.H. and Roberts, L.W. (1995). *Experiments in Plant Tissue Culture* (3rd Ed.). Cambridge: Cambridge University Press.

Duffy, E.M. and Cassells, A.C. (2003). Mycorrhizae. In Thomas, B., Murphy, D.J., and Murray, B.G. (Eds.), *Encyclopedia of Applied Plant Sciences* (pp. 1107-1115). New York: Academic Press.

Dyer, A.F. (1979). *Investigating Chromosomes*. London: Edward Arnold.

Ebbels, D.L. (2003). *Principles of Plant Health and Quarantine*. Wallingford, UK: CAB International.

Edwards, K.J. (1998). Randomly amplified polymorphic DNAs (RAPDs). In Karp, A., Isaac, P.G., and Ingram, D.S. (Eds.), *Molecular Tools for Screening Plant Diversity* (pp. 171-175). London: Chapman & Hall.

Esau, K. (1977). *Anatomy of Seed Plants*. New York: Wiley.

Faccioli, G. and Marani, F. (1998). Virus elimination by meristem tip culture and tip micrografting. In Hadidi, A., Khetarpal, R.K., and Koganezawa, H. (Eds.), *Plant Virus Disease Control* (pp. 346-380). St. Louis, MO: APS Press.

Fath, A., Bethke, P.C., and Jones, R.L. (1999). Barley aleurone cell death is not apoptotic: Characterization of nuclease activities and DNA degradation. *Plant J. 20:* 305-315.

Favier, A.E., Kalyanaraman, B., Fontcave, M., Pierre, J.L., and Cadet, E. (1995). *Analysis of Free Radicals in Biological Systems*. New York: Springer-Verlag.

Firor, J. (1992). *Changing Atmosphere: A Global Challenge*. New Haven, CT: Yale University Press.

Ford-Lloyd, B.V., Callow, A., and Newberry, H.J. (1997). *Biotechnology and Plant Genetic Resources: Conservation and Use*. Wallingford, UK: CAB International.

Foroughi-Wehr, B. and Wenzel, G. (1993). Andro- and parthenogenesis. In Hayward, M.D., Bosemark, N.O., and Romagosa, I. (Eds.), *Plant Breeding: Principles and Prospects* (pp. 261-278). London: Chapman & Hall.

Fowler, M.R. and Rayns, F.M. (1993). Core techniques for culture establishment and maintenance. In Hunter, C.S. (Ed.), *In Vitro Cultivation of Plant Cells* (pp. 19-42). Oxford: Butterworth-Heinemann.

Fox, R.T.V. (1993). *Principles of Diagnostic Techniques in Plant Pathology.* Wallingford, UK: CAB International.

Franck, T., Gaspar, T., Kevers, C., Penel, C., Dommes, J., and Hausman, J.F. (2001). Are hyperhydric shoots of *Prunus avium* L. energy deficient? *Plant Sci. 160:* 1145-1151.

Fried, B. and Sherma, J. (1999). *Thin-Layer Chromatography.* New York: Marcel Dekker.

Fujita, N. and Kinase, A. (1991). The use of robotics in automated plant propagation. In Vasil, I.K. (Ed.), *Scale-Up and Automation in Plant Propagation* (pp. 213-244). New York: Academic Press.

Fukuda H. (1996). Xylogenesis: initiation, progression and cell death. *Ann. Rev. Plant Physiol. Plant Mol. Biol. 47:* 299-325.

Fukui, K. and Nakayama, S. (1996). *Plant Chromosomes: Laboratory Methods.* Boca Raton, FL: CRC Press.

Gaba, V.P. (2005). Plant growth regulators in tissue culture and development. In Trigiano, R.N. and Gray, D.J. (Eds.), *Plant Development and Biotechnology* (pp. 87-100). Boca Raton, FL: CRC Press.

Gahan, P.B. (1972). *Autoradiography for Biologists.* London: Academic Press.

————. (1984). *Plant Histochemistry and Cytochemistry: An Introduction.* London: Academic Press.

————. (1989). Viability of plant protoplasts. In Bajaj, Y.P.S. (Ed.), *Plant Protoplasts and Genetic Engineering. Biotechnology in Agriculture and Forestry,* Vol. 8 (pp. 34-49). Berlin: Springer-Verlag.

————. (2003). Messenger DNA in higher plants. *Cell Biochem. Funct. 21:* 207-209.

Gahan. P.B., Wang, L., Bowen, I.D., and Winters, C. (2003). Cytokinin-induced apoptotic nuclear changes in cotyledons of *Solanum aviculare* and *Lycopersicon esculentum. Plant Cell Tissue Organ Cult. 72:* 237-245

Gahan, P.B. and Wyndaele, R. (2000). BAP synthesis by tomato cotyledons *in vitro* and implications for organogenesis. Abstracts COST 843 Working Group 1. Developmental Biology of Regeneration. Geisenheim, October 12-15, 2000, pp. 12-15.

Gamborg, O.L., Miller, R.A., and Ojima, K. (1968). Nutrient requirements of suspension cultures of soybean root cells. *Exp. Cell Res. 50:* 151-158.

Gaspar, T., Franck, T., Bisbis, B., Kevers, C., Jouve, L., Hausman, J.F., and Dommes, J. (2002). Concepts in plant stress physiology. Application to plant tissue cultures. *Plant Growth Regul. 37:* 263-285.

Gautheret, R.J. (1959). *La culture des tissus végétaux.* Paris: Masson et Cie.

Geier, T. and Sangwan, R.S. (1996). Histological and chimeral-segregation reveal cell-specific differences in the competence for shoot regeneration and

*Agrobacterium*-mediated transformation in *Kohleria* internode explants. *Plant Cell Rep. 15:* 386-390.

George, E.F. (1993). *Plant Propagation by Tissue Culture, Part 1: The Technology.* Basingstoke: Exegetics.

———. (1996). *Plant Propagation by Tissue Culture, Part 2: In Practice.* Basingstoke: Exegetics.

Ghannoum, M.A. and O'Toole, G.A. (2004). *Microbial Biofilms.* New Delhi: ASM Press.

Ghosh, R. (2003). *Protein Bioseparation Using Ultrafiltration: Theory, Applications and New Developments.* London: Imperial College Press.

Givan, A.L. (2001). *Flow Cytometry: First Principles.* New York: Wiley.

Glick, B.R. and Pasternak, J.J. (2003). *Molecular Biotechnology: Principles and Applications of Recombinant DNA.* Washington, DC: American Society for Microbiology.

Goel, P.K., Prasher, S.O., Landry, J.A., Patel, R.M., Bonnell, R.B., Viau, A.A., and Miller, J.R. (2003). Potential of airborne hyperspectral remote sensing to detect nitrogen deficiency and weed infestation in corn. *Comput. Electron. Agric. 38:* 99-124.

Gonzalez, R.C. and Woods, R.E. (2001). *Digital Image Processing.* Boston, MA: Addison-Wesley.

Goto, M. (1992). *Fundamentals of Bacterial Plant Pathology.* New York: Academic Press.

Graham, L.E., Graham, J.M., and Wilcox, L.W. (2003). *Plant Biology.* New York: Prentice Hall.

Grant, V. (1975). *Genetics of Flowering Plants.* New York: Columbia University Press.

Gray, D.J., Compton, M.E., Hiebert, E., Lin, C.-H., and Gaba, V.P. (2005). Construction and use of a simple gene gun for particle bombardment. In Trigiano, R.N. and Gray, D.J. (Eds.), *Plant Development and Biotechnology* (pp. 265-272). Boca Raton, FL: CRC Press.

Gray, D.J., Jayasankar, S., and Li, Z.T. (2005). A simple illumination system for visualizing green fluorescent protein. In Trigiano, R.N. and Gray, D.J. (Eds.), *Plant Development and Biotechnology* (pp. 273-276). Boca Raton, FL: CRC Press.

Grinberg, N. (2001). *Modern Thin Layer Chromatography.* New York: Marcel Dekker.

Grob, R.L. and Barry, E.F. (2004). *Modern Practice of Gas Chromatography.* New York: Wiley.

Grunewaldt-Stocker, G. (1997). Problems with plant health of *in vitro* propagated *Anthurium* spp. and *Phalaenopsis* hybrids. In Cassells, A.C. (Ed.), *Pathogen and Microbial Contamination Management in Micropropagation* (pp. 363-370). Dordrecht, The Netherlands: Kluwer Academic Publishers.

Grzebelus, E. and Adamus, A. (2004). Effect of anti-mitotic agents on the development and genome doubling of gynogenic onion (*Allium cepa* L.) embryos. *Plant Sci. 167:* 569-574.

Gunawardena, A.H.L.A.N., Pearce, M., Jackson, M.B., Hawes, C.R., and Evans, D.E. (2001) Characterization of programmed cell death during aerenchyma formation induced by ethylene or hypoxia in roots of maize (*Zea mays* L). *Planta* (Berlin) *212:* 205-214.

Gurr, G. and Wratten, S. (2002). *Biological Control: Measures of Success.* Dordrecht, The Netherlands: Kluwer Academic Publishers.

Hadidi, A., Flores, R., Randles, J.W., and Semancik, S. (2003). *Viroids.* Enfield, NH: Science Publishers.

Hadidi, A., Khetarpal, R.K., and Koganezawa, H. (1998). *Plant Virus Disease Control.* St. Louis, MO: APS Press.

Haensch, K.T. (1999). Somatic embryogenesis *in vitro* from adult plants of *Pelargonium*: Influence of genotype and basal medium. *Gartenbauwissenschaft 64:* 193-200.

Hames, B.D. and Rickwood, D. (2001). *Gel Electrophoresis of Proteins: A Practical Approach.* Oxford: Oxford University Press.

Hannon, G.J. (2003). *RNAi: A Guide to Gene Silencing.* Cold Spring Harbor, NY: Cold Spring Harbor Laboratory Press.

Harborne, J.B. and Tomas-Barberan, F.A. (1991). *Ecological Chemistry and Biochemistry of Plant Terpenoids.* Oxford: Oxford University Press.

Hari, V. and Das, P. (1998). Ultra microscopic detection of plant viruses and their gene products. In Hadidi, A., Khetarpal, R.K., and Koganezawa, K. (Eds.), *Plant Virus Disease Control* (pp. 417-427). St. Louis, MO: APS Press.

Hartmann, H.T., Davies, F.T., Kester, D.E., and Geneve, R.L. (2001). *Hartmann and Kester's Plant Propagation: Principles and Practice.* Harlow, UK: Pearson Education.

Hauptmann, R.M., Widholm, J.M., and Paston, J.D. (1985). Benomyl: A broad spectrum fungicide for use in plant cell and protoplast culture. *Plant Cell Rep. 4:* 129-132.

Hawes, C.R. and Satiat-Jeunemaitre, B. (2001). *Plant Cell Biology: A Practical Approach.* Oxford: Oxford University Press.

Heller, R. (1953). Researches on the mineral nutrition of plant tissues. *Ann. Sci. Nat. Bot. Biol. Veg. 14:* 1-223.

———. (1955). Les besoins mineraux des tissues en culture. *Union Int. Sci. Biol. Series B 20:* 1-21.

Hemingway, R.W. and Karchesy, J.J. (1989). *Chemistry and Significance of Condensed Tannins.* Boulder, CO: Perseus.

Hemingway, R.W. and Laks, P.E. (1992). *Plant Polyphenols: Synthesis, Properties and Significance.* New York: Plenum Publishing.

Hendry, G.A.F. (1993). Plant pigments. In Lea, P.J. and Leegood, R.C. (Eds.), *Plant Biochemistry and Molecular Biology* (pp. 181-196). New York: Wiley.

Herzberg, O., Frankel, H., Brown, A.H.D., and Burdon, J.J. (1995). *Conservation of Plant Diversity.* Cambridge: Cambridge University Press.

Hewitt, H.G. (1998). *Fungicides in Crop Protection.* Wallingford, UK: CAB International.

Heywood, V.H. (1993). *Flowering Plants of the World*. Oxford: Oxford University Press.

Hoagland, D.R. (1948). *Lectures on Inorganic Nutrition of Plants*. Waltham, MA: Chronica Botanica.

Hoagland, D.R. and Arnon, D.I. (1938) . The water culture method for growing plants without soil. *Univ. Calif. Agric. Exp. Sta. Circ.* 347.

Holst, G.C. (2000). *Common Sense Approach to Thermal Imaging*. Bellingham, WA: International Society for Optical Engineering.

Hopfenberg, H.B. and Stannett, V. (1974). *Permeability of Plastic Films and Coatings to Gases, Vapours and Liquids*. Boulder, CO: Perseus Publishing.

Hopkins, W.G. and Hunter, N.P. (2002). *Introduction to Plant Physiology*. New York: Wiley.

Howell, S.H. (1998). *Molecular Genetics of Plant Development*. Cambridge: Cambridge University Press.

Hull, R. (2002). *Matthews' Plant Virology*. San Diego, CA: Academic Press.

Hunter, C.S. (1993). Protoplasts and haploid cultures. In Hunter, C.S. (Ed.), *In Vitro Cultivation of Plant Cells* (pp. 87-112). Oxford: Butterworth-Heinemann.

Hvoslef-Eide, T. and Preil, W. (2004). *Liquid Culture Systems for In Vitro Mass Propagation of Plants*. Dordrecht, The Netherlands: Springer.

Imeson, A. (1999). *Thickening and Gelling Agents for Food*. London: Blackie.

Inze, D. and Van Montague, M. (2001). *Oxidative Stress in Plants*. London: Taylor & Francis.

Jayasankar, S. and Gray, D.J. (2005). Variation in tissue culture. In Trigiano, R.N. and Gray, D.J. (Eds.), *Plant Development and Biotechnology* (pp. 301-310). Boca Raton, FL: CRC Press.

Joyce, S.M. and Cassells, A.C. (2002). Variation in potato microplant morphology *in vitro* and DNA methylation. *Plant Cell Tissue Organ Cult. 70:* 125-137.

Joyce, S.M., Cassells, A.C., and Mohan Jain, S. (2003). Stress and aberrant phenotypes *in vitro* culture. *Plant Cell Tissue Organ Cult 74:* 103-121.

Jucker, E. (2001). *Antiviral Agents: Advances and Problems*. Basel: Birkhauser Verlag.

Kane, M.E. (2005). Shoot culture procedures. In Trigiano, R.N. and Gray, D.J. (Eds.), *Plant Development and Biotechnology* (pp. 145-158). Boca Raton, FL: CRC Press.

Kao, K.N. and Michayluk, M.R. (1975). Nutrient requirements for the growth of *Vicia hajastana* cells and protoplasts at very low population density in liquid media. *Planta 126:* 105-110.

Kapulnik, Y. and Douds, D.D. (2000). *Arbuscular Mycorrhizas: Physiology and Function*. Dordrecht, The Netherlands: Kluwer Academic Publishers.

Karp, A. (1990). Somaclonal variation in potato. In Bajaj, Y.P.S. (Ed.), *Biotechnology in Agriculture and Forestry, Vol. 11. Somaclonal Variation in Crop Improvement I* (pp. 379-399). Berlin: Springer-Verlag.

Karp, A., Isaac, P.G., and Ingram, D.S. (1998). *Molecular Tools for Screening Biodiversity*. London: Chapman & Hall.

Kasukabe, Y., He, L.X., Nada, K., Misawa, S., Ihara, I., and Tachibana, S. (2004). Overexpression of spermidine synthase enhances tolerance to multiple environmental stresses and up-regulates the expression of various stress regulated genes in transgenic *Arabidopsis thaliana*. *Plant Cell Physiol. 45:* 712-722.

Kaufman, S.C., Musch, D.C., Belin, M.W., Cohen, E.J., Meisler, D.M., Reinhart, W.J., Udell, I.J., and Van Meer, W.S. (2004). Confocal microscopy. *Ophthalmology 111:* 396-406.

Kevers, C., Franck, T., Strasser, R.J., Dommes, J., and Gaspar, T. (2004). Hyperhydricity of micropropagated shoots: A typically stress-induced change of physiological state. *Plant Cell Tissue Organ Cult. 77:* 181-191.

Kiersteins, G. (1994). Effects of low light intensity and high air humidity on morphology and permeability of plant cuticles, with special respect to plants cultured *in vitro*. In Lumsden, P.J., Nicholas, J.R., and Davies, W.J. (Eds.), *Physiology, Growth and Development of Plants in Culture* (pp. 132-142). Dordrecht, The Netherlands: Kluwer Academic Publishers.

King, E.O., Ward, M.K., and Raney, D.E. (1954). Two simple media for the demonstration of pyocyanin and fluorescein. *J. Lab. Clin. Med. 44:* 301-307.

Kipling, D. (1996). *Telomere*. Oxford: Oxford University Press.

Klerks, M.M., Leone, G.O.M., van den Heuvel, J.F.J.M., and Schoen, C.D. (2001). Development of a multiplex AmplidDet RNA for the simultaneous detection of potato leaf roll virus and potato virus Y in potato tubers. *J. Virol. Methods 93:* 115-125.

Knudson, L. (1946). A new nutrient solution for the germination of orchid seed. *Am. Orchid. Soc. 15:* 214-217.

Kosslak, R.M., Chamberlin, M.A., Palmer, R.G., and Bowen, B.A. (1997). Programmed cell death in the cortex of soybean root necrosis mutants. *Plant J. 11:* 729-745.

Kozai, T. (1991). Controlled environments in conventional and automated micropropagation. In Vasil, I.K. (Ed.), *Scale-Up and Automation in Plant Propagation* (pp. 213-230). New York: Academic Press.

Krczal, G. (1998). Virus certification of ornamental plants—The European strategy. In Hadidi, A., Khetarpal, R.K., and Koganezawa, H. (Eds.), *Plant Virus Disease Control* (pp. 277-287). St. Louis, MO: APS Press.

Kuriyama, H. and Fukuda, H. (2002). Developmental programmed cell death in plants. *Curr. Opin. Plant Biol. 5:* 568-573.

Kutter, E. and Sulakvelidze, A. (2004). *Bacteriophages*. Boca Raton, FL: CRC Press.

Laimer, M. and Rucker, W. (2003). *Plant Tissue Culture*. New York: Springer-Verlag.

Larsen, A.H., Feddersen, R., and Palsboll, P.J. (1998). A rapid screening procedure for detecting mtDNA halophytes in humpback whales (*Megaptera novaeangliae*). In Karp, A., Isaac, P.G., and Ingram, D.S. (Eds.), Molecular Tools for Screening Biodiversity (pp. 157-163). London: Chapman & Hall.

Lawson, E.J.R. and Poethig, R.S. (1995). Shoot development in plants: Time for a change. *Trends Genet. 11*, 263-268.

Leifert, C. and Cassells, A.C. (2001). Microbial hazards in plant tissue and cell cultures. *In Vitro Plant 37:* 133-138.

Leifert, C. and Waites, W.M. (1992). Bacterial growth in plant tissue cultures. *J. Appl. Bacteriol. 72:* 460-466.

Leigh, R.A. and Sanders, D. (1998). *Plant Vacuole,* Vol. 25. New York: Elsevier.

Lelliot, R.A. and Stead, D.E. (1987). *Methods for the Diagnosis of Bacterial Diseases of Plants.* Oxford: Blackwell.

Lengeler, J.W., Schlegel, H.G., and Drews, G. (1999). *Biology of the Prokaryotes.* Oxford: Blackwell.

Lerner, H.R. (1999). *Plant Responses to Environmental Stresses.* New York: Marcel Dekker.

Lewis, T. (1998). *Thrips as Crop Pests.* Wallingford, UK: CAB International.

Li, Z.T. and Gray, D.N. (2005). Genetic engineering technologies. In Trigiano, R.N. and Gray, D.J. (Eds.), *Plant Development and Biotechnology* (pp. 241-250). Boca Raton, FL: CRC Press.

Liebler, D.C. (2002). *Introduction to Proteomics: Tools for the New Biology.* Totowa, NJ: Humana Press.

Linsdey, K. and Yeoman, M.M. (1983). Novel experimental systems for studying the production of secondary metabolites by plant tissue culture. In Mantell, S.H. and Smith, H. (Eds.), *Plant Biotechnology, Soc. Exp. Biol. Seminar Series* 18 (pp. 39-66). Cambridge: Cambridge University Press.

Linsmaier, E.M. and Skoog, F. (1965). Organic growth factor requirements of tobacco tissue cultures. *Plant Physiol. 18:* 100-127.

Lloyd, G. and McCowan, B. (1981). Commercially-feasible micropropagation of mountain laurel, Kalmia latifolia, by use of shoot tip culture. *Int. Plant Prop. Soc. Proc. 30:* 421-427.

Long, R.D. (1997). Photoautotrophic micropropagation—a strategy for contamination control. In Cassells, A.C. (Ed.), *Pathogen and Microbial Contamination Management in Micropropagation* (pp. 267-278). Dordrecht, The Netherlands: Kluwer Academic Publishers.

Lopez-Delgado, H., Mora-Herrera, M.E., Zavaleta-Mancera, H.A., Cadena-Hinojosa, M., and Scott, I.M. (2004). Salicylic acid enhances heat tolerance and potato virus X (PVX) elimination during thermotherapy of potato microplants. *Am. J. Pot. Res. 81:* 171-176.

Lumsden, P.J., Nicholas, J.R., and Davies, W.J. (1994). *Physiology, Growth and Development of Plants in Culture.* Dordrecht, The Netherlands: Kluwer Academic Publishers.

Lyndon, R.F., Barlow, P.W., Kirk, D.L., and Bard, J.B.L. (1998). *The Shoot Apical Meristem: Its Growth and Development.* Cambridge: Cambridge University Press.

Maene, L. and Debergh, P. (1985). Liquid medium additions to established tissue cultures to improve elongation and rooting *in vivo. Plant Cell Tissue Organ Cult. 5:* 23-33.

Margulis, L. (1981). *Symbiosis in Cell Evolution.* San Francisco, CA: W.H. Freeman & Co.

Marschner, H. (1994). *Mineral Nutrition of Higher Plants*. London: Academic Press.

Martienssen, R., Volpe, T., Lippman, Z., Gendrel, A.-V., Kidner, C., Rabinowicz, P., and Colot, V. (2004). Transposable elements, RNA interference and the origin of heterochromatin. In Hannon, G.J. (Ed.), *RNAi: A Guide to Gene Silencing* (pp. 129-148). Cold Spring Harbor, NY: Cold Spring Harbor Laboratory Press.

Martin, R.R. (1998). Advanced diagnostic tools as an aid to controlling plant virus diseases. In Hadidi, A., Khetarpal, R.K., and Koganezawa, H. (Eds.), *Plant Virus Disease Control* (pp. 381-391). St. Louis, MO: APS Press.

Massey, L.K. (2002). *Permeability of Plastics and Elastomers: A Guide to Packaging and Barrier Materials*. New York: William Andrew.

Matthes, M.C., Daly, A., and Edwards, K.J. (1998). Amplified fragment length polymorphism (AFLP). In Karp, A., Isaac, P.G., and Ingram, D.S. (Eds.), *Molecular Tools for Screening Biodiversity* (pp. 183-190). London: Chapman & Hall.

Matthews, G.A. (1999). *Applications of Pesticides to Crops*. London: Imperial College Press.

Matzke, M.A. and Matzke, A.J.M. (2000). *Plant Gene Silencing*. Dordrecht, The Netherlands: Kluwer Academic Publishers.

Mauseth, J.D. (1988). *Plant Anatomy*. San Francisco, CA: Benjamin-Cummings.

———. (1998). *Botany*. Sudbury, MA: Jones and Bartlett.

———. (2003). *Botany: Introduction to Plant Biology* (3rd Ed.). Sudbury, MA: Jones and Bartlett.

McCabe, P.F., Levine, A., Meijer, P.-J., Tapon, N.A., and Penel, R.I. (1997). A programmed cell death pathway activated in carrot cells cultured at low density. *Plant J. 12:* 267-280.

McClintock, B. (1951). Chromosome organization and genic expression. *Cold Spr. Harb. Symp. 21:* 197-216.

McMaster, M.C. (1994). *HPLC: A Practical User's Guide*. New York: Wiley.

McMillan Browse, P. (1979). *Plant Propagation*. London: Mitchell Beazley.

McNair, H.M. and Miller, J.M. (1997). *Basic Gas Chromatography*. New York: Wiley.

McPherson, M.J., Moller, S., Howe, C.D., and Beynon, R.J. (2000). *PCR Basics*. New York: Springer-Verlag.

Menard, D., Coumans, M., and Gaspar, T. (1985). Micropropagation du *Pelargonium* a partir de meristemes. *Med. Fac. Landbouww. Rijksuniv. Gent. 50:* 237-331.

Metzler, D.E. (2001). *Biochemistry*, Vol. 1 (2nd Ed.). New York: Academic Press.

———. (2003). *Biochemistry*, Vol. 2 (2nd Ed.). New York: Academic Press.

Meulemans, M., Duchene, D., and Fouarge, G. (1987). Selection of variants by dual culture of potato and *Phytophthora infestans*. In Bajaj, Y.P.S. (Ed.), *Biotechnology in Agriculture and Forestry*, Vol. 3 (pp. 318-331). Berlin: Springer-Verlag.

Micke, A. and Donini, B. (1993). Induced mutations. In Hayward, M.D., Bosemark, N.O., and Romagosa, I. (Eds.), *Plant Breeding: Principles and Prospects* (pp. 52-62). London: Chapman & Hall.

Mihaljevic, S., Peric, M., and Jelska, S. (2001). The sensitivity of embryonic tissues of *Picea omorika* (Panc.) Purk. to antibiotics. *Plant Cell Tissue Organ Cult. 67:* 287-293.

Mink, G.I. (1998). Virus certification of deciduous fruit trees in the United States and Canada. In Hadidi, A., Khetarpal, R.K., and Koganezawa, H. (Eds.), *Plant Virus Disease Control* (pp. 294-300). St. Louis, MO: APS Press.

Mink, G.I., Wample, R., and Howell, W.E. (1998). Heat treatment of perennial plants to eliminate phytoplasmas, viruses and viroids while maintaining plant survival. In Hadidi, A., Khetarpal, R.K., and Koganezawa, H. (Eds.), *Plant Virus Disease Control* (pp. 332-345). St. Louis, MO: APS Press.

Mittler, R. and Shulaev, V. (2003). Apoptosis in plants, yeast and bacteria. In Yin, X.-M. and Dong, Z. (Eds.), *Essentials of Apoptosis* (pp. 125-133). Totowa, NJ: Humana Press.

Mohan Jain, S., Brar, D.S., and Ahloowalia, B.S. (1998). *Somaclonal Variation and Induced Mutations in Crop Improvement*. Dordrecht, The Netherlands: Kluwer Academic Publishers.

Mohan Jain, S. and Ishii, K. (2003). *Micropropagation of Woody Trees and Fruits*. Dordrecht, The Netherlands: Kluwer Academic Publishers.

Moore, T.C. (1989). *Biochemistry and Physiology of Plant Hormones*. Berlin: Springer-Verlag.

Mortimore, S. and Wallace, C. (2001). *HACCP: A Practical Approach*. Biggleswade, UK: Aspen Law and Business.

Motulsky, H. (1995). *Intuitive Biostatistics*. Oxford: Oxford University Press.

Mount, D.W. (2002). *Bioinformatics: Sequence and Genome Analysis*, Vol. 5. Cold Spring Harbor, NY: Cold Spring Harbor Laboratory Press.

Mukerji, K.G. and Upadhyay, K. (1999). *Biotechnological Approaches in Biocontrol of Plant Pathogens*. Dordrecht, The Netherlands: Kluwer Academic Publishers.

Murashige, T. and Skoog, F. (1962). A revised medium for the rapid growth and bio-assays with tobacco tissue cultures. *Physiol. Plant. 15:* 473-497.

Murphy, D.B. (2001). *Fundamentals of Light Microscopy and Electronic Imaging*. New York: Wiley.

Murphy, D.J. (1993). Plant lipids: Their metabolism, functions and utilization. In Lea, P.J. and Leegood, R.C. (Eds.), *Plant Biochemistry and Molecular Biology* (pp. 113-128). New York: Wiley.

Nassar, A.H., El-Tarabily, K.A., and Sivasithamparam, K. (2003). Growth promotion of bean (*Phaseolus vulgaris* L.) by a polyamine-producing isolate of *Streptomyces griseoluteus. Plant Growth Regul. 40:* 97-106.

Neogi, P. (1996). *Diffusion in Polymers*. New York: Marcel Dekker.

Nierhaus, K.H. and Wilson, D.N. (2004). *Protein Synthesis and Ribosome Structure: Translating the Genome*. New York: Wiley.

Nilson, E.T. and Orcutt, D.M. (1996). *The Physiology of Plants under Stress, Vol. 1. Environmental Factors*. New York: Wiley.

Nitsch, C. (1977). Culture of microspores. In Reinert, J. and Bajaj, Y.P.S. (Eds.), *Applied and Fundamental Aspects of Plant Cell, Tissue and Organ Culture* (pp. 268-278). Berlin: Springer-Verlag.

Nitsch, J.P. and Nitsch, C. (1969). Haploid plants from pollen grains. *Science 163:* 85-87.

Nogué, F., Gonneau, M., and Faure, J.-D. (2003). Cytokinins. In Henry, H.L. and Norman, A.W. (Eds.), *Encyclopaedia of Hormones* (pp. 371-378). New York: Academic Press.

Novina, C.D. and Sharp, P.A. (2004). The RNAi revolution. *Nature 430:* 161-164.

Nowak, J., Asiedu, S.K., Bensalim, S., Richards, J., Stewart, A., Smith, A.C., Stevens, D., and Sturz, A.V. (1997). From laboratory to applications: Challenges and progress with *in vitro* dual cultures of potato and beneficial bacteria. In Cassells, A.C. (Ed.), *Pathogen and Contamination Management in Micropropagation* (pp. 321-330). Dordrecht, The Netherlands: Kluwer Academic Publishers.

O'Herlihy, E.A. and Cassells, A.C. (2003). Influence of *in vitro* factors on titre and elimination of model fruit tree viruses. *Plant Cell Tissue Organ Cult. 72:* 33-42.

Orcutt, D.M. and Nilsen, E.T. (2000). *Physiology of Plants under Stress: Soil and Biotic Factors,* Vol. 2. New York: Wiley.

Ormerod, M.G. (2004). Cell-cycle analysis of asynchronous populations. In Hawley, T.S. and Hawley, R.G. (Eds.), *Methods in Molecular Biology, Vol. 263. Flow Cytometry Protocols* (2nd Ed.) (pp. 345-354). Totowa, NJ: Humana Press.

Ortiz-Ortiz, L., Bojalil, L.F., and Yakoleff, V. (1984). *Actinomycetes Biology.* Amsterdam: Elsevier.

Osborn, K.R. and Jenkins, W. (1992). *Plastic Films.* Boca Raton, FL: CRC Press.

Parekh, S.R. and Vinci, V.A. (2003). *Handbook of Industrial Cell Culture: Mammalian, Microbial and Plant Cells.* Totowa, NJ: Humana Press.

Petroski, R.J., Behle, R., and Tellez, M.R. (2004). *Semiochemicals in Pest Management and Alternative Agriculture.* Oxford: Oxford University Press.

Pierik, R.L. (2002). *In Vitro Culture of Higher Plants.* Dordrecht, The Netherlands: Kluwer Academic Publishers.

Poethig, R.S. (1990). Phase change and the regulation of shoot morphogenesis. *Science 250:* 923-930.

———. (2003). Phase change and the regulation of developmental timing in plants. *Science 301:* 334-336.

Poincare, H., Schuepp, H., Hasselwandter, K., and Barea, J.M. (2002). *Mycorrhizal Technology in Agriculture: From Genes to Bioproducts.* Basel: Birkhauser Verlag.

Postgate, J.R. (1998). *Nitrogen Fixation.* Cambridge: Cambridge University Press.

Prasad, K.N. (1995). *Handbook of Radiation Biology.* Boca Raton, FL: CRC Press.

Preece, J.E. and Sutter, E.G. (1991). Acclimatization of microplants to the greenhouse and field. In Debergh, P.C. and Zimmerman, R.H. (Eds.), *Micropropagation* (pp. 71-93). New York: Academic Press.

Prescott, L.M., Klein, D.A., and Harley, J.P. (2001). *Microbiology.* New York: McGraw-Hill.

Pype, J., Everaert, K., and Debergh, P. (1997). Contamination by micro-arthropods in plant tissue culture. In Cassells, A.C. (Ed.), *Pathogen and Microbial Contamination Management in Micropropagation* (pp. 259-266). Dordrecht, The Netherlands: Kluwer Academic Publishers.

Quicke, D.L.J. (1993). *Principles and Techniques of Contemporary Taxonomy.* Oxford: Blackie.

Quorin, M. and Lepoivre, P. (1977). Improved media for *in vitro* culture of *Prunus* sp. *Acta Hortic. 78:* 437-442.

Raghavendra, A.S. (1998). *Photosynthesis: A Comprehensive Treatise.* Cambridge: Cambridge University Press.

Rangaswamy, N.S. (1977). Applications of *in vitro* pollination and *in vitro* fertilization. In Reinert, J. and Bajaj, Y.P.S. (Eds.), *Plan Cell, Tissue and Organ Culture* (pp. 412-425). Berlin: Springer-Verlag.

Rao, K. (1999). *Photosynthesis.* Cambridge: Cambridge University Press.

Rayns, F.W. and Fowler, M.R. (1993). Media design and use. In Hunter, C.F. (Ed.), *In Vitro Cultivation of Plant Cells* (pp. 43-64). Oxford: Butterworth-Heinemann.

Razdan, M.K. (2002). *Introduction to Plant Tissue Culture.* Enfield, NH: Science Publishers.

Redenbaugh, V.J. (1993). *Synseeds: Application of Synthetic Seeds to Crop Improvement.* Boca Raton, FL: CRC Press.

Reed, S.M. (2005). Haploid cultures. In Trigiano, R.N., and Gray, D.J. (Eds.), *Plant Development and Biotechnology* (pp. 225-234). Boca Raton, FL: CRC Press.

Resh, H.M. (2002). *Hydroponic Food Production: A Definitive Guide to Soilless Culture.* Erlbaum Associates, Mahwah, NJ: Woodbridge Press.

Ridge, I. (2002). *Plants.* Oxford: Oxford University Press.

Rival, A., Jaligot, E., Beule, T., Verdeil, J.-L., and Tregear, J. (2000). DNA methylation and somaclonal variation in oil palm. *Acta Hortic. 530:* 447-454.

Roberts, J., Downs, S., and Parker, P. (2002). Plant growth and development. In Ridge, I. (Ed.), *Plants* (pp. 221-274). Oxford: Oxford University Press.

Roberts, J.A. and Tucker, G.A. (1999). *Plant Hormone Protocols*, Vol. 141. Totowa, NJ: Humana Press.

Roberts, M.F. and Wink, M. (1998). *Alkaloids: Biochemistry, Ecology and Medicinal Applications.* Dordrecht, The Netherlands: Kluwer Academic Publishers.

Roberts, T.R. (1998). *Metabolic Pathways of Agrochemicals: Herbicides and Plant Growth.* London: Royal Society of Chemistry.

Roe, R.M., Kuhr, R.J., and Burton, J.D. (1997). *Herbicide Activity: Toxicology, Biochemistry and Molecular Biology.* Amsterdam: IOS Press.

Roh, K.S. and Choi, B.Y. (2004). Sucrose regulates growth and activation of rubisco in tobacco leaves *in vitro. Biotechnol. Bioprocess Eng. 9:* 229-239.

Roitt, I., Male, D.K., and Brostoff, J. (2001). *Immunology.* Amsterdam: Elsevier Science.

Rose, J. (2002). *The Plant Cell Wall.* Boca Raton, FL: CRC Press.

Russell, P.J. (2002). *Genetics.* San Francisco, CA: Benjamin Cummings.

Russo, V.E., Salamini, F., Edgar, L.G., Jaenisch, R., and Cove, D.J. (1999). *Development: Genetics, Epigenetics and Environmental Regulation.* New York: Springer-Verlag.

Ryder, M.H., Stephens, P.M., and Bowen, G.D. (1995). *Improving Plant Productivity with Rhizosphere Bacteria.* Collingwood, Victoria, Australia: CSIRO.

Satina, S., Blakeslee, A.F., and Avery, A.G. (1940). Demonstration of the three germ layers in the shoot apex of *Datura* by means of induced polyploidy in periclinal chimeras. *Am. J. Bot.* 27: 895-905.

Saxena, A.K. and Johri, B.N. (2002). *Arbuscular Mycorrhizae: Interactions in Plants, Rhizosphere and Soil.* Enfield, NH: Science Publishers.

Schaad, N.W., Jones J.B., and Chun, W. (2001). *Laboratory Guide for Identification of Plant Pathogenic Bacteria.* St. Louis, MO: APS Press.

Schena, M. (2004). *Protein Microarrays.* Sudbury, MA: Jones and Bartlett.

Schepin, O. and Ermakov, W. (1991). *International Quarantine.* Guilford, CT: International University Press.

Schulz, S. (2004). *Chemistry of Pheromones and Other Semiochemicals 1.* New York: Springer-Verlag.

Schwarz, M. (1995). *Soilless Culture Management.* New York: Springer-Verlag.

Schwarz, O.J., Sharma, A.R., and Beaty, R.M. (2005). Propagation from nonmeristematic tissues: Organogenesis. In Trigiano, R.N. and Gray, D.J. (Eds.), *Plant Development and Biotechnology* (pp. 159-172). Boca Raton, FL: CRC Press.

Scott, G. (1997). *Antioxidants: In Science, Technology, Medicine and Nutrition.* New York: Horwood Publishing.

Scragg, A. (1991a). Plant cell bioreactors. In Stafford, A. and Warren, G. (Eds.), *Plant Cell and Tissue Culture* (pp. 221-239). Milton Keynes, UK: Open University Press.

————. (1991b). The immobilization of plant cells. In Stafford, A. and Warren, G. (Eds.), *Plant Cell and Tissue Culture* (pp. 205-220). Milton Keynes, UK: Open University Press.

Shalitin l.C. (2003). Cryptochrome structure and signal transduction. *Annu. Rev. Plant Physiol. Plant Mol. Biol.* 54: 469-496.

Sharma, A.K. and Sharma, A. (1999). *Plant Chromosomes: Analysis, Manipulation, Engineering.* New York: Taylor & Francis.

Sheppard, C.J.R., Shotton, D., and Hotton, D.M. (1997). *Confocal Laser Scanning Microscopy.* Abingdon, UK: BIOS Scientific.

Sherma, J. and Fried, B. (2003). *Handbook of Thin-Layer Chromatography.* New York: Marcel Dekker.

Sholto Douglas, J. (1986). *Advanced Guide to Hydroponics.* London: Pelham Books.

Sigee, D.C. (1993). *Bacterial Plant Pathology: Cell and Molecular Aspects.* Cambridge: Cambridge University Press.

Singh, R.P. and Dhar, A.K. (1998). Detection and management of plant viroids. In Hadidi, A., Khetarpal, R.K., and Koganezawa, H. (Eds.), *Plant Virus Disease Control* (pp. 428-447). St. Louis, MO: APS Press.

Singhal, G.S., Sopory, S.K., Renger, G., Irrgang, K.D., and Govindjee, K.Y. (1999). *Concepts in Photobiology: Photosynthesis and Photomorphogenesis.* Dordrecht, The Netherlands: Kluwer Academic Publishers.

Skoog, F. and Miller, C.O. (1957). Chemical regulation of growth and organ formation in plant tissues cultured *in vitro. Symp. Soc. Exp. Bot. 11:* 118-131.

Slater, A., Scott, N.W., and Fowler, M.R. (2003). *Plant Biotechnology.* Oxford: Oxford University Press.

Slocum, R.D. and Flores, H.E. (1991). *Biochemistry and Physiology of Polyamines in Plants.* Boca Raton, FL: CRC Press.

Smith, C.J. (1993). Carbohydrate chemistry. In Lea, P.J. and Leegood, R.C. (Eds.), *Plant Biochemistry and Molecular Biology* New York: Wiley.

Smith, R.H. (2000). *Plant Tissue Culture: Techniques and Experiments* (2nd Ed.). New York: Academic Press.

Soll, D., Moore, P., and Nishimura, S. (2001). *RNA.* Amsterdam: Elsevier.

Spirin, A.S. (2000). *Ribosomes.* Dordrecht, The Netherlands: Kluwer Academic Publishers.

Srivastava, L.M. (2002). *Plant Growth and Development.* New York: Elsevier Science.

Stafford, A. and Warren, G. (1991). *Plant Cell and Tissue Culture.* Milton Keynes, UK: Open University Press.

Stead, D.E., Elphinstone, J.G., Weller, S., Smith, N., and Hennessy, J. (2000). Modern methods for characterising, identifying and detecting bacteria associated with plants. *Acta Hortic. 530:* 45-60.

Stead, D.E., Sellwood, J.E., Wilson, J., and Viney, I. (1992). Evaluation of a commercial microbial identification system based on fatty acid profiles for rapid, accurate identification of plant pathogenic bacteria. *J. Appl. Biochem. 72:* 315-321.

Stepan-Sarkissian, G. (1991). Biotransformation of plant cell cultures. In Stafford, A. and Warren, G. (Eds.), *Plant Cell and Tissue Culture* (pp. 163-204). Milton Keynes, UK: Open University Press.

Strange, R. (2003). *Introduction to Plant Pathology.* New York: Wiley.

Streibig, J.C. and Kudsk, P. (1992). *Herbicide Bioassays.* Boca Raton, FL: CRC Press.

Suttle, G.R.L. (2005). Commercial laboratory production. In Trigiano, R.N. and Gray, D.J. (Eds.), *Plant Development and Biotechnology* (pp. 311-319). Boca Raton, FL: CRC Press.

Swartz, H.J. (1991). Post culture behaviour: Genetic and epigenetic effects and related problems. In Debergh, P.C. and Zimmerman, R.H. (Eds.), *Micropropagation: Technology and Applications* (pp. 95-122). Dordrecht, The Netherlands: Kluwer Academic Publishers.

Taiz, L. and Zeigler, E. (2002). *Plant Physiology* (3rd Ed.). Sunderland, MA: Sinauer Associates.

Taji, A., Kumar, P.P., and Lakshmanan, P. (2001). *In Vitro Plant Breeding.* Binghamton, NY: Haworth Press.

Takayama, S., Swedlund, B., and Miwa, Y. (1991). Automated propagation of microbulbs of lilies. *Cell Cult. Somatic Cell Genet. Plants. 8:* 111-131.

Tate, R.L. (2000). *Soil Microbiology.* New York: Wiley.

Thompson, M. Haeusler, R.A., Good, P.D., and Engelke, D.R. (2003) Nucleolar clustering of dispersed tRNA genes. *Science 302:* 1399-1401.

Thorpe, T.A. (2002). *In Vitro Embryogenesis in Plants.* Dordrecht, The Netherlands: Kluwer Academic Publishers.

Tilney-Bassett, R.A. (1991). *Plant Chimeras.* Cambridge: Cambridge University Press.

Tosi, R. Luigetti, R., and Zazzerini, A. (1999). Benzothiadiazole induces resistance to *Plasmopara helianthi* in sunflower plants. *J. Phytopath. 147:* 365.

Towers, G.H. and Stafford, H.A. (1990). *Biochemistry of the Mevalonic Acid Pathway to Terpenoids.* New York: Plenum Press.

Towill, L.E. (2005). Germplasm conservation. In Trigiano, R.N. and Gray, D.J. (Eds.), *Plant Development and Biotechnology* (pp. 277-284). Boca Raton, FL: CRC Press.

Tran Thanh Van, M. (1973). *In vitro* control of *de novo* flower, bud, root and callus differentiation from excised epidermal tissues. *Nature 246:* 44-45.

Tran Thanh Van, M., Dien, N.T., and Chlyaha (1974). Regulation of organogenesis in small explants of superficial tissue of *Nicotiana tabacum* L. *Planta 119:* 149-159.

Trigiano, R.N. and Gray, D.J. (2000). *Plant Tissue Culture Concepts and Laboratory Exercises* (2nd Ed.). Boca Raton, FL: CRC Press.

————. (2005). *Plant Development and Biotechnology.* Boca Raton, FL: CRC Press.

Trigiano, R.N., Malueg, K.R., Pickens, K.A., Cheng, Z.-M., and Graham, E.T. (2005). In Trigiano, R.N. and Gray, D.J. (Eds.), *Plant Development and Biotechnology* (pp. 39-48). Boca Raton, FL: CRC Press.

Tulecke, W., Weinstein, L.H., Rutner, A., and Laurencot, H.J. (1961). The biochemical composition of coconut milk as related to its use in plant tissue culture. *Contrib. Boyce Thompson Inst. 21:* 115-128.

Turner, B.S. (2001). *Chromatin and Gene Regulation.* Oxford: Blackwell Publishers.

Tyree, M.T. and Zimmermann, M.H. (2002). *Xylem Structure and the Ascent of Sap.* New York: Springer-Verlag.

Van de Vijver, G. and de Waale, D. (2002). *From Epigenesis to Epigenetics: The Genome in Context.* New York: New York Academy of Sciences.

Van der Greef, J., Stroobant, P., and van der Heijden, R. (2004). The role of analytical sciences in medical systems biology. *Curr. Opin. Chem. Biol. 8:* 1-7.

Van der Linde, P.C.G. (2000). Certified plants from tissue culture. *Acta Hortic. 530:* 93-102.

Van Driesche, R.G. and Bellows, T.S. (1996). *Biological Control.* London: Chapman & Hall.

Van Harten, A.M. (1998). *Mutation Breeding: Theory and Practical Applications.* Cambridge: Cambridge University Press.

Van Loon, L.C. (2000). Systemic induced resistance. In Slusarenko, A., Fraser, R.S.S., and van Loon, L.C. (Eds.), *Mechanisms of Resistance to Plant Diseases* (pp. 521-574). Dordrecht, The Netherlands: Kluwer Academic Publishers.

Van Tuyl, J.M., Lim, K.B., and Ramanna, M.S. (2002). Interspecific hybridisation and introgression. In Vainstein, A. (Ed.), *Breeding for Ornamentals: Classical and Molecular Approaches* (pp. 85-103). Dordrecht, The Netherlands: Kluwer Academic Publishers.

Vasil, I.K. (1991). *Scale-Up and Automation in Micropropagation.* San Diego, CA: Academic Press.

Veilleux, R.E., Compton, M.E., and Saunders, J.A. (2005). Use of protoplasts. In Trigiano, R.N. and Gray, D.J. (Eds.), *Plant Development and Biotechnology* (pp. 213-224). Boca Raton, FL: CRC Press.

Vestberg, M., Cassells, A.C., Schubert, A., Cordier, C., and Gianinazzi, S. (2002). AMF and micropropagation of high value crops. Chapter 16. In Gianinazzi S., Schüepp H., Barea J.M., and Haselwandter K. (Eds.), *Mycorrhizal Technology in Agriculture: From Genes to Bioproducts* (pp. 223-233). Zurich: Birkhäuser

Vickery, M.L. and Vickery, B. (1981). *Secondary Plant Metabolism.* London: Macmillan.

Walker, G.M. (1998). *Yeast Physiology and Biotechnology.* New York: Wiley.

Wallsgrove, R.M. (1995). *Amino Acids and Their Derivatives in Higher Plants.* Cambridge: Cambridge University Press.

Walsh, C. (2003). *Antibiotics: Actions, Origins, Resistance.* Herndon, VA: ASM Press.

Walters, D.R. (2003). Polyamines and plant disease. *Phytochemistry 64:* 97-107.

Wang, M., Heokstra, S., van Bergen, S., Lamers, G.E.M., Oppedijk, B.J., van der Hiejden, M.W., de Priester, W., and Schilperoort, R.A. (1999). Apoptosis in developing anthers and the role of ABA in this process during androgenesis in *Hordeum vulgare* L. *Plant. Mol. Biol. 39:* 489-503.

Warren, G. (1991). Protoplast isolation and fusion. In Stafford, A. and Warren, G. (Eds.), *Plant Cell and Tissue Culture* (pp. 48-81). Milton Keynes, UK: Open University Press.

Watkinson, S.C., and Gooday, G.W. (2001). *The Fungi.* London: Academic Press.

Weier, T.E., Stocking, C.R., and Barbour, M.G. (1974). *Botany: An Introduction to Plant Biology* (5th Ed.), New York: Wiley.

Wendehenne, D., Durner, J., and Klessig, D.F. (2004). Nitric oxide: A new player in plant signalling and defence. *Curr. Opin. Plant Biol. 7:* 1-7.

Westermeier, R. and Barnes, N. (2001). *Electrophoresis in Practice.* New York: Wiley.

Westhoff, P. (1998). *Molecular Plant Development: From Gene to Plant.* Oxford: Oxford University Press.

Weston, L.A. and Duke, S.O. (2003). Weed and crop allelopathy. *Crit. Revs. Plant Sci. 22:* 367-389.

Williamson, B., Cooke, D.E.L., Duncan, J.M., Leifert, C., Breese, W.A., and Shattock, R.C. (1997). Fungal infections of micropropagated plants at weaning: A problem exemplified by downy mildews in *Rubus* and *Rosa.* In Cassells, A.C.

(Ed.), *Pathogen and Contamination Management in Micropropagation* (pp. 309-320). Dordrecht, The Netherlands: Kluwer Academic Publishers.

Willis, K.J. and McElwain, J.C. (2002). *The Evolution of Plants.* Oxford: Oxford University Press.

Winfrey, M.R., Rott, M.A., and Wortman, A.T. (1997). *Unravelling DNA: Molecular Biology for the Laboratory.* New York: Simon & Schuster.

Wink, B. (1999). *Biochemistry of Plant Secondary Metabolism.* Boca Raton, FL: CRC Press.

Winsor, G.W. and Schwarz, M. (1990). *Soilless Culture for Horticultural Crop Production.* Lanham, MD: Bernan Associates.

Wooding, F.B.P. and Head, J.J. (1978). *Phloem.* Burlington, NC: Carolina Biological Supply Company.

Yokoda, T. (1997). The structure, biosynthesis and function of brassinosteroids. *Trends Plant Sci. 2:* 137-143.

Ziv, M. (1991). Vitrification: Morphological and physiological disorders of *in vitro* plants. In Debergh, P.C. and Zimmerman, R.H. (Eds.), *Micropropagation: Technology and Application* (pp. 45-70). Dordrecht, The Netherlands: Kluwer Academic Publishers.

Ziv, M. and Ariel, T. (1994). Vitrification in relation to stomatal deformation and malfunction in carnation leaves *in vitro.* In Lumsden, P.J., Nicholas, J.R., and Davies, W.J. (Eds.), *Physiology, Growth and Development of Plants in Culture* (pp. 143-154). Dordrecht, The Netherlands: Kluwer Academic Publishers.

Zobayed, S.M.A., Afreen, F., and Kozai, T. (2000). Quality biomass production via photoautotrophic micropropagation. *Acta Hortic. 530:* 377-386.

Zvereva, M.E., Shpanchenko, O.V., Dontsova, O.A., and Bogdanov, A.A. (2000). Structure and function of tmRNA (10Sa RNA). *Mol. Biol. 34:* 927-933.